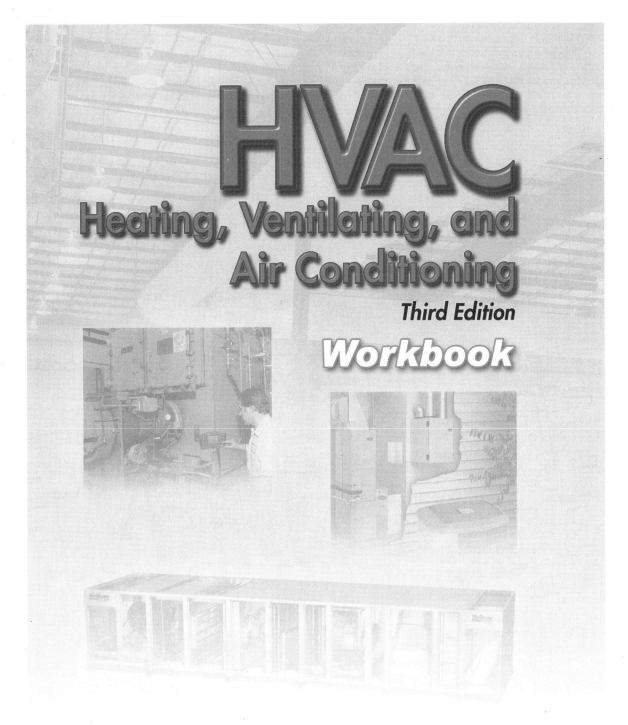

HVAC
Heating, Ventilating, and Air Conditioning
Third Edition
Workbook

AMERICAN TECHNICAL PUBLISHERS, INC.
HOMEWOOD, ILLINOIS 60430-4600

S. Don Swenson

Heating, Ventilating, and Air Conditioning Workbook contains HVAC procedures commonly practiced in industry and the trade. Specific procedures vary with each task and must be performed by a qualified person. For maximum safety, always refer to specific manufacturer recommendations, insurance regulations, specific job site and plant procedures, applicable federal, state, and local regulations, and any authority having jurisdiction. The material contained is intended to be an educational resource for the user. American Technical Publishers, Inc. assumes no responsibility or liability in connection with this material or its use by any individual or organization.

Aquastat is a registered trademark of Honeywell Heating Specialties Company. Pyrex is a registered trademark of Corning Incorporated.

3 4 5 6 7 8 9 – 04 – 9 8 7 6 5

Printed in the United States of America

ISBN 978-0-8269-0679-3

Contents

Introduction

Review Questions, Activities, and Trade Tests in *HVAC Workbook* should be completed after studying the corresponding chapter in *Heating, Ventilating, and Air Conditioning*. Specific instructions are provided throughout this workbook. Charts and tables required to complete various Activities and Trade Tests in *HVAC Workbook* are included on pages 171–202.

Question formats include True-False, Multiple Choice, Completion, Matching, and Calculations. Always record the answer in the space(s) provided. All answers are given in the *HVAC Answer Key*.

True-False

Circle T if the statement is true. Circle F if the statement is false.

T (F) **1.** Hydrogen and oxygen are the two elements found in all fuels.

(T) F **2.** Hydrogen burns faster than carbon.

Multiple Choice

Select the response that correctly completes the statement. Write the appropriate letter in the space provided.

B **1.** In the combustion process _____ and oxygen combine.
  A. heat   C. air
  B. fuel   D. fire

Completion

Determine the response that correctly completes the statement. Write the appropriate response in the space provided.

belt **1.** Two blower drives used on furnaces are direct and _____.

drum **2.** Two heat exchangers used in furnaces are clam shell and _____.

Matching

Select the correct response. Write the appropriate letter in the space provided.

C **1.** Atmospheric air   **A.** Air without moisture

A **2.** Dry air   **B.** Dry air and moisture

B **3.** Moist air   **C.** Air and moisture mixture

Calculations

Solve the problem. Write the appropriate answer in the space provided.

600,000 Btu **1.** If a consumer uses 6 therms of natural gas in a given period of time, how much heat is produced?

182,500 Btu **2.** How much heat is produced when 73 cu ft of propane is burned at 100% efficiency?

TEST FORMATS

Name _Thomas O'Connell_   Date _1/24/2012_

True-False

(T) F   **1.** An air circulation rate of 40 fpm is considered ideal.

(T) F   **2.** A relative humidity of 50% feels comfortable to most people.

T **(F)**   **3.** Building related illness and sick building syndrome are the same thing.

T **(F)**   **4.** Air velocity does not affect the rate of evaporation of moisture.

(T) F   **5.** Air velocity is the speed at which air moves from one place to another.

(T) F   **6.** The presence of pollutant sources is an important factor influencing indoor air quality.

(T) F   **7.** Pesticides used for pest management in building spaces can be a very serious pollutant source if not used correctly.

(T) F   **8.** Relative humidity is always expressed as a percentage.

T **(F)**   **9.** The higher the relative humidity, the faster the rate of evaporation.

T **(F)**   **10.** Comfort is the condition that occurs when people can sense a difference between themselves and the surrounding air.

(T) F   **11.** Air in a building must be circulated to provide maximum comfort.

(T) F   **12.** Stagnation occurs when there is no circulation of air.

(T) F   **13.** Humidity affects the rate of evaporation of perspiration from the body.

T **(F)**   **14.** Fresh air is introduced into the return air portion of an HVAC system. True

(T) F   **15.** The amount of ventilation air necessary for comfort is determined by the number of occupants and the kind of activity taking place in a building.

Completion

humidity   **1.** _____ is the amount of moisture in the air.

relative   **2.** _____ humidity is the amount of moisture in the air compared to the amount of moisture the air holds when saturated.

Stagnant   **3.** _____ air lacks oxygen required to provide comfort.

filters   **4.** The air in a forced-air system is cleaned by _____ placed in return air ductwork.

BTU   **5.** A(n) _____ is the quantity of heat required to raise 1 lb of water 1°F.

dewpoint   **6.** _____ is the temperature below which moisture begins to condense.

ton of cooling   **7.** A(n) _____ is the amount of cooling required to melt a ton of ice in a 24-hour period.

humidifier   **8.** A(n) _____ adds moisture to air by causing water to evaporate.

Comfort   **9.** _____ is the condition that occurs when a person cannot sense a difference between themselves and the surrounding air.

ventilation **10.** _____ is the process of introducing fresh air into a building.

Multiple Choice

D.   **1.** Air velocity is expressed in _____.
- A. gallons per hour
- B. feet per hour
- C. gallons per minute
- D. feet per minute

A.   **2.** For maximum comfort, temperature controls must maintain air temperature within _____°F or _____°F of a comfortable temperature.
- A. 1; 2
- B. 2; 4
- C. 3; 6
- D. 5; 10

C.   **3.** A ton of cooling is equal to _____ Btu/hr.
- A. 12
- B. 1200
- C. 12,000
- D. 120,000

A.   **4.** The type, location, and _____ of registers determine the amount of supply air that is introduced into each building space.
- A. size
- B. cost
- C. style
- D. manufacturer

B.   **5.** A _____ circulates air in an HVAC system.
- A. furnace
- B. blower
- C. humidifier
- D. filter

C.   **6.** A(n) _____ removes particulate matter from the air.
- A. dehumidifier
- B. air conditioner
- C. filter
- D. blower

B.   **7.** A dehumidifier _____ air.
- A. adds moisture to
- B. removes moisture from
- C. circulates
- D. cleans

D.   **8.** Temperature stratification is a result of poor _____.
- A. ventilation
- B. cooling
- C. heating
- D. circulation

D.   **9.** Common reasons for poor indoor air quality in building spaces are _____.
- A. air pollutant sources
- B. poor ventilation
- C. unanticipated building uses
- D. all of the above

C.   **10.** People feel most comfortable when the relative humidity is _____%.
- A. 10
- B. 30
- C. 50
- D. 70

Name _____   Date _____

Activity — Air Conditions

Select a room or building for a series of measurements.

1. Walk through the building and record your feelings of comfort.

2. Explain the conditions that caused the level of comfort in the building.

3. Measure the temperature and relative humidity in the building. Check the ventilation system to see if filtered fresh air is being introduced. Estimate the velocity of the air moving through the building.

   A. Temperature _____°F

   B. Relative humidity _____%

   C. Fresh air _____

   D. Air velocity _____fpm

4. What conclusions can be drawn about the HVAC system?

Activity — Practical Application

In some applications, comfort conditions are overridden by health considerations. Place an **H** in front of the applications where health is of most importance. Place a **C** in front of the applications where comfort is of most importance.

_____ 1. Hospital   _____ 4. Indoor athletic event

_____ 2. Office   _____ 5. Nursing home

_____ 3. Clinic   _____ 6. Residence

Activity — Identification

Identify the components of a forced-air system that are related to comfort. Record the name of the part and the function it provides in a total comfort system.

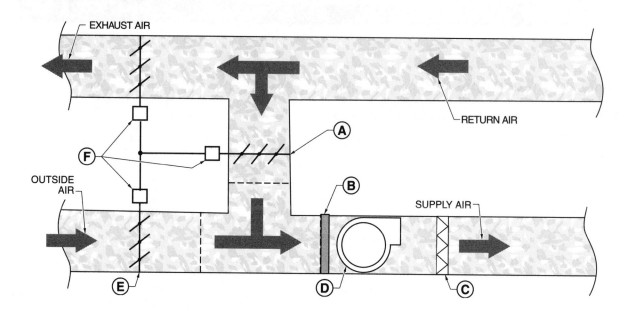

A. Name

Function

B. Name

Function

C. Name

Function

D. Name

Function

E. Name

Function

F. Name

Function

Name _____ Date _____

Completion

_____ 1. People feel most comfortable when the temperature in a room is _____°F.

_____ 2. People feel most comfortable when the air velocity is _____ fpm.

_____ 3. Humidity is the amount of _____ in the air.

_____ 4. _____ is the process of introducing fresh air into a building.

_____ 5. A(n) _____ removes moisture from the air.

_____ 6. _____ is the movement of air.

_____ 7. _____ is the process that occurs when a liquid changes to a vapor by absorbing heat.

_____ 8. One ton of cooling equals _____ Btu per 24-hour period.

_____ 9. _____ air is used to replace air that is lost to exhaust.

_____ 10. Normal body temperature is _____°F.

True-False

T   F   1. HVAC systems use equipment consisting of mechanical, pneumatic, electrical, electronic, and chemical components.

T   F   2. Humidity is always present in the air.

T   F   3. Filters circulate air in a forced-air system.

T   F   4. Health effects of poor indoor air quality can show up after a single exposure.

T   F   5. Body temperature of a person is controlled to some extent by perspiration.

T   F   6. The design, maintenance, and operation of a ventilation system is only a minor factor in poor indoor air quality.

T   F   7. Regardless of the temperature of the air, air moving through a room at a velocity of 65 fpm feels drafty.

T   F   8. An increase in air velocity reduces the rate of evaporation of perspiration from the skin and makes a person feel warm.

T   F   9. Physiological factors are adjustments made by the human body in response to surrounding conditions.

T   F   10. Ammonia is a source of pollution that is found in the glues in furniture, in carpeting, and in a host of other building materials.

T F **11.** High humidity indicates dry air that contains little moisture.

T F **12.** Temperature stratification occurs when warm air in a building space rises to the ceiling and cold air drops to the floor.

T F **13.** Improperly located outdoor air intakes can bring in air contaminants such as vehicle exhaust, combustion emissions, and other types of pollution.

T F **14.** Condensation is the process that occurs when a liquid changes to a vapor by absorbing heat.

T F **15.** A dehumidifier cools air.

T F **16.** A great way to control indoor air pollution is to eliminate furniture, carpeting, and other furnishings that emit formaldehyde and other volatile organic compounds.

T F **17.** Temperature control equipment maintains the temperature in a building within 1°F or 2°F of a setpoint temperature for maximum comfort.

T F **18.** The amount of moisture required to saturate the air changes as the dry bulb temperature changes.

T F **19.** Moisture is never present in the air.

T F **20.** Humidity is not important in determining comfort.

T F **21.** With no temperature change and a decrease in relative humidity, a person feels warmer because of the slower evaporation rate.

T F **22.** In buildings with swimming pools or a large number of potted plants, the humidity level may be too low for comfort.

T F **23.** All filters are designed for disposal after use.

T F **24.** Ventilation and circulation are necessary for comfort inside a building.

T F **25.** Repeated exposures to poor indoor air quality may be required before health effects show up.

Matching

_____ **1.** Stratification

_____ **2.** Stagnation

_____ **3.** Ventilation

_____ **4.** Filtration

_____ **5.** Circulation

A. Air cleaning
B. Air movement
C. Temperature layers
D. Still air
E. Outdoor air

Name _____ Date _____

True-False

T	F	**1.**	A standard is a regulation or minimum requirement.
T	F	**2.**	The danger signal word indicates an imminently hazardous situation which will result in death or serious injury.
T	F	**3.**	A Class B protective helmet is required in utility service for high-voltage protection.
T	F	**4.**	The severity of hearing loss depends on the intensity and duration of exposure.
T	F	**5.**	Tagout is the process of removing the source of electrical power and installing a lock that prevents the power from being turned ON.
T	F	**6.**	Solid-state devices and circuits may be damaged or destroyed by a 10 V electrostatic discharge.
T	F	**7.**	A harness is a rubber, leather, or plastic pad strapped onto the knees for protection.
T	F	**8.**	Safety nets are optional when public traffic or other workers are permitted underneath a work area where they are not otherwise protected from falling objects.
T	F	**9.**	A danger tag may be used alone when a lock does not fit the disconnect device.
T	F	**10.**	A lockout/tagout may be removed by anyone who needs to operate the equipment.

Completion

_____ **1.** A(n) _____ is a person who has special knowledge, training, and experience in the installation, programming, maintenance, and troubleshooting of HVAC equipment.

_____ **2.** The _____ signal word indicates a potentially hazardous situation which could result in death or serious injury.

_____ **3.** _____ is gear worn by HVAC or maintenance technicians to reduce the possibility of injury in the work area.

_____ **4.** Safety glasses are an eye protection device with special _____ glass or plastic lenses.

_____ **5.** A(n) _____ is made of rope or webbing for catching and protecting a falling worker.

_____ **6.** Ear protection includes earplugs and _____.

_____ **7.** The _____ signal word indicates a potentially hazardous situation which may result in minor or moderate injury.

_____ **8.** Use only a Class _____ fire extinguisher on electrical fires.

_____ **9.** _____ is securely connecting a harness directly or indirectly to an overhead anchor point.

_____ **10.** _____ are lightweight enclosures that allow the lockout of standard control devices.

Multiple Choice

_____ **1.** The three most common signal words are _____.
   A. danger, warning, careful   C. danger, warning, caution
   B. dangerous, warning, caution   D. dangerous, warning, careful

_____ **2.** The _____ warning symbol and signal word indicates a high-voltage location and condition that could result in death or serious personal injury from an electric shock.
   A. explosion   C. caution
   B. electrical   D. all of the above

_____ **3.** A _____ is a rope or webbing that is attached to a worker and tie-off device to prevent the worker from hitting the ground during a fall.
   A. lifeline   C. lanyard
   B. safety net   D. rope grab

_____ **4.** A safety net must be used where personnel are _____ feet or more above ground where the worker is not otherwise protected by a harness, lifeline, or scaffold.
   A. 6   C. 25
   B. 12   D. 50

_____ **5.** _____ are eye and face protection that cover the entire face with a plastic shield.
   A. safety glasses   C. face shields
   B. goggles   D. all of the above

_____ **6.** _____ is the connection of all exposed non-current-carrying metal parts to the earth.
   A. Short circuit   C. Lockout
   B. Electric shock   D. Grounding

_____ **7.** A(n) _____ is a printed document used to relay hazardous material information from the manufacturer, importer, or distributor to the technician.
   A. material safety data sheet   C. accident report
   B. safety label   D. fire department notification

_____ **8.** _____ forms provide a written record of the steps taken to comply with industrial safety standards as well as a method of tracking progress.
   A. Lockout/tagout   C. Accident report
   B. Documentation   D. MSDS

_____ **9.** A(n) _____ is a rope or webbing device that connects a harness to a lifeline.
   A. rope grab   C. buckle strap
   B. safety net   D. lanyard

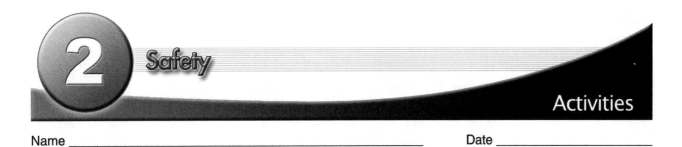

Name _____ Date _____

Activity — Inspection

After obtaining the proper authority, inspect the HVAC equipment on a construction site, large building, or manufacturing complex. List all potential equipment hazards and the safety practices required.

Date _____ Site _____

Person in charge _____ Permission granted by _____

Hazards

Safety practices

Activity — Signal Word Identification

List the signal word and significance for each symbol.

A. Signal word:

    Significance:

B. Signal word:

    Significance:

C. Signal word:

    Significance:

RED	⚠	SIGNAL WORD SIGNIFICANCE
ORANGE	⚠	SIGNAL WORD SIGNIFICANCE
YELLOW	⚠	SIGNAL WORD SIGNIFICANCE

Activity – Fall Protection Equipment Identification

Identify the fall protection equipment. Record the name of the equipment and the function it serves.

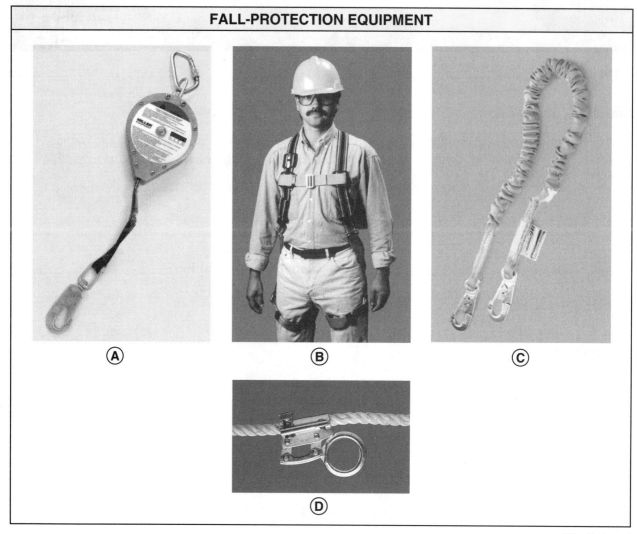

FALL-PROTECTION EQUIPMENT

Miller Equipment

A. Name:

Function:

C. Name:

Function:

B. Name:

Function:

D. Name:

Function:

2 Safety

Name _____ Date _____

True-False

T   F   **1.** Personal protective equipment (PPE) is gear worn by HVAC or maintenance technicians to reduce the possibility of injury in the work area.

T   F   **2.** The danger signal word indicates a potentially hazardous situation which could result in death or serious injury.

T   F   **3.** Ear protection includes earplugs and earmuffs.

T   F   **4.** An accident report is a printed document used to relay hazardous material information from the manufacturer, importer, or distributor to the technician.

T   F   **5.** A Class C fire extinguisher is to be used for electrical fires.

Completion

_____   **1.** A(n) _____ is a regulation or minimum requirement.

_____   **2.** The intensity of hearing loss depends on the _____ and duration of exposure.

_____   **3.** A safety net is made of rope or webbing for _____ and protecting a falling worker.

_____   **4.** Tying off is securely connecting a _____ directly or indirectly to an overhead anchor point.

_____   **5.** Ground is the connection of all exposed non-current-carrying metal parts to the _____.

Multiple Choice

_____   **1.** _____ is the process of removing the source of electrical power and installing a lock that prevents the power from being turned ON.
    A. Tagout   C. Grounding
    B. Insulation   D. Lockout

_____   **2.** The _____ signal word indicates a potentially hazardous situation which may result in minor or moderate injury.
    A. explosion   C. danger
    B. warning   D. caution

_____ **3.** A _____ must be used where personnel are 25′ or more above ground where the worker is not otherwise protected by a harness, lifeline, or scaffold.

    A. danger signal   C. protective helmet
    B. safety net   D. lockout device

_____ **4.** Face shields are eye and face protection that cover the _____ with a plastic shield.

    A. eyes only   C. entire face
    B. eyes and ears only   D. none of the above

_____ **5.** A lockout device is a lightweight enclosure that allows the _____ of standard control devices.

    A. tagout   C. lockout
    B. grounding   D. labeling

Name _Thomas O'Connell_   Date _1/24/2012_

True-False

combustions

T (F) **1.** Heat is produced by conduction, electrical energy, and alternate heat sources.

T (F) **2.** A radiant heating system heats the air in a building directly by radiation.

(T) F **3.** Resistance to the flow of electricity produces heat.

(T) F **4.** Alternate heat sources do not use combustion or electricity as sources of heat.

(T) F **5.** Heat is transferred by conduction, convection, and radiation.

T (F) **6.** Temperature is the measurement of the quantity of heat.

(T) F **7.** Heat is energy identified by a temperature difference or a change of state.

T (F) **8.** The three methods of heat transfer are convection, combustion, and radiation.

convection

(T) (F) **9.** Conduction is heat transfer that occurs when currents circulate between warm and cool regions of a fluid.

T (F) **10.** The second law of thermodynamics states that heat always flows from a material at a low temperature to a material at a high temperature.

Multiple Choice

_C._ **1.** Latent heat is heat identified by a change of _____ of a material.
  A. shape   C. state
  B. humidity   D. temperature

_A._ **2.** _____ heat is the ability of a substance to carry heat.
  A. Sensible   C. Specific
  B. Latent   D. Radiant

_A._ **3.** A temperature of 72°F equals _____°C. $\dfrac{72-32}{1.8} = 22.2\overline{2}$
  A. $22.2\overline{2}$ C. 57.7
  B. 40   D. 187.2

_B._ **4.** _____ is heat transfer that occurs when molecules in a material are heated and the heat is passed from molecule to molecule through the material.
  A. Thermal radiation   C. Convection
  B. Conduction   D. Combustion

_C._ **5.** The combustion process requires fuel, oxygen, and _____.
  A. air   C. heat
  B. radiation   D. conduction

Matching

G	1. Ignition temperature	A.	Light cannot pass through
C	2. 1 Btu	B.	Heat in the earth
F	3. Temperature	C.	Quantity of heat
A	4. Opaque	D.	Heat transfer by currents
I	5. Magma	E.	Heat transfer from molecule to molecule
E	6. Conduction	F.	Intensity of heat
D	7. Convection	G.	Heat required for combustion
J	8. Radiation	H.	1°F per pound of water
B	9. Geothermal	I.	Molten rock
H	10. Btu	J.	Heat transfer by waves

Short Answer

1. Explain the difference between heat and temperature.

   Temperature measures the intensity of heat. Heat is simply the form of energy identified by a change of state. Temp. measures the energy.

2. Define temperature difference.

   The difference between the temp of two materials. It must exist between two materials or at least two locations within a material.

3. List and describe the three methods by which heat is transferred.

   Convection: currents - usually air or water transport heat. Conduction: molecules speed up and heat spreads from one to another. radiation: travels by waves and loses heat until it hits an object and is absorbed.

4. Explain the difference between a conductivity (k) factor and a conductance (C) factor.

   Conductivity is the amount of heat transfered through 1 sq. ft of material that is 1" thick Btu p. hour, per 1°F difference. Conductance applies to the surface of an area.

5. Define thermodynamics.

   Is the science of heat and how it transforms to and from other forms of energy.

Name _____   Date _____

Activity — Temperature

Obtain a small cardboard box with a lid. Suspend a light bulb in the box. Be sure the bulb does not touch the box. Suspend a thermometer through the lid. *Note:* Use a low-wattage bulb.

1. Record the initial temperature in the box and then turn the light ON. Record the temperature at 1 minute intervals for 10 minutes.

2. On the graph, plot temperature increase against time.

Time (in minutes)	Temperature (in °F)
Initial	
1	
2	
3	
4	
5	
6	
7	
8	
9	
10	

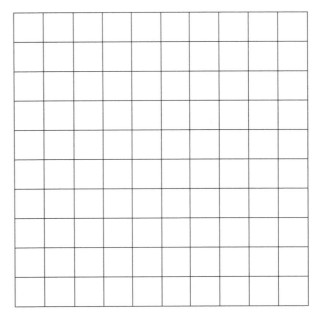

TEMPERATURE (in °F)

TIME (in minutes)

3. Calculate the rate of temperature increase in degrees Fahrenheit per minute.

4. How is the electricity used by the light bulb related to the heat recorded by temperature increase?

Activity — Flame Heat

Attach the sensors of an electronic thermometer at each end of an iron bar. Suspend the iron bar so that one end is in the upper part of the flame from a Bunsen burner. Light the burner.

1. Record the initial temperature of the bar. Record the temperature of each end of the bar at 1 minute intervals for 10 minutes.

2. On the graph, plot the temperature increase of each end of the bar.

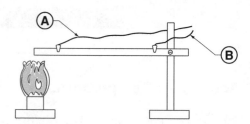

Time (in minutes)	Temperature (in °F)	
	A	**B**
Initial		
1		
2		
3		
4		
5		
6		
7		
8		
9		
10		

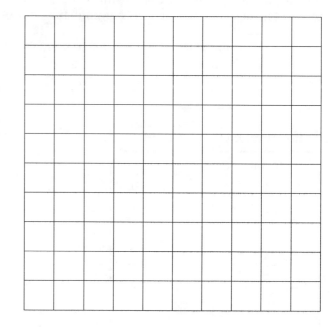

TEMPERATURE (in °F)

TIME (in minutes)

3. How does the heat travel along the bar?

4. How long does it take for the heat to travel from one end of the bar to the other?

5. When the experiment is continued, is the temperature of the bar the same along its entire length?

Activity — Furnace

Use any furnace that can be operated in a normal heating mode. Cycle the furnace ON. Measure and record the temperature of the air entering the furnace by placing a thermometer in the return air plenum.

_____   **1.** The return air temperature is _____°F.

At the same time, measure and record the temperature of the air leaving the furnace by placing a thermometer in the supply air plenum. *Note:* The thermometer placed in the supply air plenum must be able to measure high temperatures.

_____   **2.** The supply air temperature is _____°F.

3. What does the temperature of the air entering the furnace indicate about the heat content of the air? Explain.

4. What does the temperature of the air leaving the furnace indicate about the heat content of the air? Explain.

5. What is the relationship between the heat content of the air entering and the air leaving the furnace? Explain.

6. What value(s) is/are needed to determine the amount of heat added to the air to cause the temperature increase?

7. What is the temperature increase of the air?

Activity — Specific Heat

Place a beaker containing 1 lb of water on a heat source. One pint of water weighs about 1 lb.

1. Measure and record the temperature of the water before the heat is turned ON.

2. Turn the burner ON. Record the temperature of the water at 1 minute intervals. Record the time elapsed and the temperature at each time interval on the graph. Shut the burner OFF when the water begins to boil. Connect the points on the graph to obtain a curve that shows how heat flows to the water from the burner.

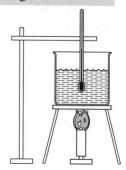

Time (in minutes)	Temperature (in °F)
Initial	
1	
2	
3	
4	
5	
6	
7	
8	
9	
10	

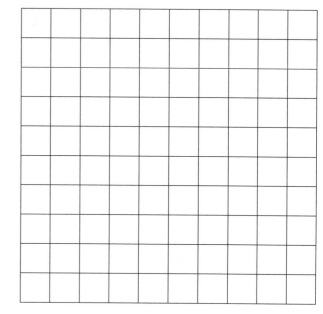

TIME (in minutes)

3. If the weight of water and the temperature increase are known, can the amount of heat added to the water be calculated? Explain.

3 Thermodynamics and Heat

Trade Test

Name _____   Date _____

Calculations

_____ **1.** Calculate the heat required to raise the temperature of 125 lb of water 10°F.

_____ **2.** Concrete 8″ thick has a C factor of .90. The heat conducted through 47 sq ft of the concrete if the temperature difference through it is 70°F is _____ Btu/hr.

_____ **3.** Gypsum board ½″ thick has a C factor of 2.22. The amount of heat conducted through 96 sq ft of gypsum board if the temperature difference through the board is 44°F is _____ Btu/hr.

_____ **4.** The specific heat of air at standard conditions is .24 Btu/lb. The amount of heat required to change the temperature of 462 lb of air from 45°F to 72°F is _____ Btu.

_____ **5.** The temperature of 1200 lb of air is 42°F as it is brought into a building. How much heat is required to raise the air temperature to 72°F?

_____ **6.** The specific heat of aluminum is .214 Btu/lb. The amount of heat required to raise 2 lb of aluminum 80°F is _____ Btu.

_____ **7.** The wall of a room is 16′ long and 8′ high. The surface area of the wall is _____ sq ft.

_____ **8.** Glass has a k factor of .35. The amount of heat transferred through 1 sq ft of glass 1″ thick with a 30°F temperature difference through the glass is _____ Btu/hr.

_____ **9.** A 20″ long drum heat exchanger has a diameter of 12″. Find the surface area of the heat exchanger.

_____ **10.** The specific heat of water is 1.0 Btu/lb. The amount of heat required to raise 4.5 lb of water 67°F is _____ Btu.

_____ **11.** The surface area of an aluminum heat exchanger is 16 sq ft. The k factor for aluminum is 1536.0. Find the amount of heat transferred through the heat exchanger if an 80°F temperature difference exists on each side of the exchanger.

_____ **12.** The specific heat of iron is .12 Btu/lb. The amount of heat required to raise 146 lb of iron 10°F is _____ Btu.

_____ **13.** A piece of sheet metal is 24″ long and 16″ wide. Find the surface area.

_____ **14.** The hardwood floor of a residence has a surface area of 600 sq ft. The k factor for hardwood is 1.10. The amount of heat transferred through the wood if a 15°F temperature difference exists on both sides of the wood is _____ Btu/hr.

_____ **15.** A 3′ long steel drum has a diameter of 2′. Find the surface area of the drum.

Completion

_____ **1.** _____°C equals 65°F.

_____ **2.** _____°C equals −34°F.

_____ **3.** _____°C equals −21°F.

_____ **4.** _____°F equals 46°C.

_____ **5.** _____°F equals 23°C.

Matching

_____ **1.** 42.8°F

_____ **2.** 5°C

_____ **3.** 57.2°F

_____ **4.** 12°C

_____ **5.** 212°F

_____ **6.** 158°F

_____ **7.** 0°C

_____ **8.** 35°C

_____ **9.** 68°F

_____ **10.** 25°C

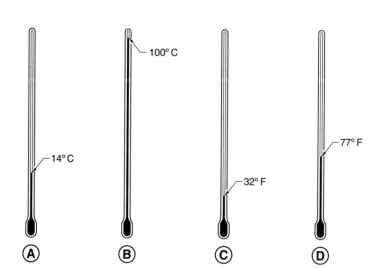

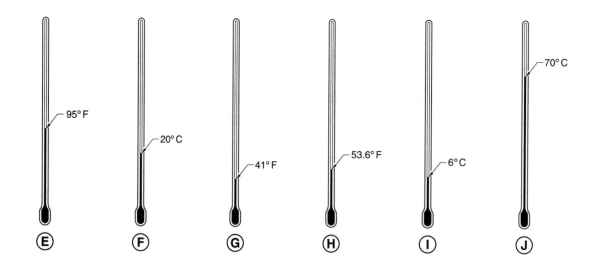

Combustion and Fuels

Review Questions

Name _____ Date _____

Completion

_____ 1. _____ is the solid fuel used most commonly for heating.

_____ 2. _____ occurs if there is not enough air supplied for combustion.

_____ 3. _____ is the most commonly used gas fuel for heating.

_____ 4. _____ coal yields a large amount of volatile matter when it burns.

_____ 5. _____ is the ability of a liquid or semiliquid to resist flow.

_____ 6. _____ are combustible materials other than coal, oil, and natural gas.

_____ 7. _____ is unburned particles of carbon that are carried away from the flame by the convective currents generated by the heat of the flame.

_____ 8. Crude oil is separated into petroleum products at a(n) _____.

_____ 9. _____ is air provided at a burner for proper combustion of fuel.

_____ 10. _____ is a chemical reaction where compounds break apart into elements and release chemical energy in the form of light and heat.

_____ 11. Combustion is the chemical reaction of _____, hydrogen, and carbon.

_____ 12. _____ are classified by physical characteristics or chemical composition.

_____ 13. Heating value is expressed in _____ for liquid fuels.

_____ 14. Grades of fuel oil are refined from _____.

_____ 15. When combustion efficiency is _____, most of the chemical energy is converted to thermal energy.

_____ 16. Fossil fuels include coal, _____, and natural gas.

_____ 17. A(n) _____ of gas fuel produces 100,000 Btu of heat.

_____ 18. Fuel, oxygen, and the proper _____ are requirements for combustion.

_____ 19. _____ coal yields a small amount of volatile matter.

_____ 20. _____ produces carbon dioxide and water vapor.

True-False

T   F   1. Hydrogen and oxygen are the two elements found in all fuels.

T   F   2. Hydrogen burns faster than carbon.

T   F   3. Bituminous coal is a hard, high luster coal that yields a small amount of volatile matter when burned.

T   F   **4.** The amount of air required for complete combustion depends on the type of fuel used.

T   F   **5.** Hydrogen passes through the combustion process unchanged.

T   F   **6.** LP gas is a mixture of propane, butane, and manufactured gas.

T   F   **7.** Coal is classified according to grade and weight.

T   F   **8.** Of all gas fuels, natural gas has the highest heating value.

T   F   **9.** Hydrocarbons are compounds that contain hydrogen and carbon.

T   F   **10.** All fuels have the same heating value.

T   F   **11.** The outer mantle of a flame is produced when hydrogen reacts with oxygen.

T   F   **12.** The inner mantle of a flame is light blue in color.

T   F   **13.** Natural gas requires approximately 135 cu ft of air for every cubic foot of gas burned.

T   F   **14.** Incomplete combustion produces carbon monoxide, aldehydes, and water vapor.

T   F   **15.** A stack thermometer is a thermometer used to measure high flue-gas temperatures.

Multiple Choice

_____   **1.** In the combustion process _____ and oxygen combine.

     A. heat   C. air
     B. fuel   D. fire

_____   **2.** Atmospheric air contains approximately _____% oxygen.

     A. 15.4   C. 20.95
     B. 16.54   D. 29.5

_____   **3.** Grades of fuel oil are based on viscosity and _____.

     A. thickness   C. color
     B. weight   D. none of the above

_____   **4.** _____ passes through the combustion process unchanged.

     A. Oxygen   C. Nitrogen
     B. Carbon   D. Hydrogen

_____   **5.** The air-fuel ratio for natural gas must be between _____% and _____% for combustion to take place.

     A. 4; 14   C. 20; 30
     B. 10; 20   D. 35; 45

4 Combustion and Fuels

Name _____ Date _____

Activity — Combustion Air

Light a Bunsen burner. Adjust the air shutter on the burner so the flame burns cleanly.

1. Where is the air mixing with the fuel?

Readjust the air shutter on the burner to make the flame yellow. Shut OFF the air at the bottom

2. Where is the air mixing with the fuel

3. The two types of flame represent what two types of combustion?

4. Why does the flame appear yellow when the air to the burner tube is shut OFF?

Activity — Combustion

Light a candle and observe combustion.

1. What is the fuel for combustion?

2. What is the source of the ignition temperature?

3. What is the source of the oxygen for combustion?

Activity — Kindling Temperature

Attempt to light a candle using a glowing splinter of wood, sparks from a welding sparklighter, and a burning match.

1. Which attempt was the most successful?

2. How was each attempt to light the candle different?

3. If the fuel source is a gas fuel, is there a different result from each attempt?

Activity — Flame Parts

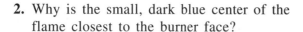

Light a Bunsen burner or a propane hand torch and adjust the flame for normal operation.

1. Identify the parts of the flame.

    A. _____

    B. _____

    C. _____

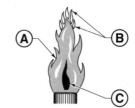

2. Why is the small, dark blue center of the flame closest to the burner face?

3. What chemical element in fuel causes the light blue part of the flame?

Name _____ Date _____

Calculations

_____ 1. If a consumer uses 6 therms of natural gas in a given period of time, how much heat is produced?

_____ 2. How much heat is produced when 73 cu ft of propane is burned at 100% efficiency?

_____ 3. How much heat is produced when 215 lb of coal is burned at 100% efficiency? Use the standard heating value for coal.

_____ 4. How much combustion air is required to burn 1.25 gal. of oil in an oil-burning furnace?

_____ 5. If 5 kW of electrical energy is converted to heat, how much heat is produced?

_____ 6. The heat produced when 16 cu ft of natural gas is burned at 100% efficiency is _____ Btu.

_____ 7. How much heat is produced when 8 gal. of Grade No. 2 fuel oil is burned at 100% efficiency? Use the standard heating value for Grade No. 2.

_____ 8. How much heat is produced when 1.5 therms of natural gas is burned?

_____ 9. The amount of air required to burn 20 lb of coal is _____ cu ft.

_____ 10. How much heat is produced from 18.6 kW of electrical energy?

_____ 11. How much heat is produced when .3 gal. of Grade No. 2 fuel oil is burned at 100% efficiency? Use the standard heating value for Grade No. 2.

_____ 12. How many therms of natural gas is required to produce 370,000 Btu?

_____ 13. How much heat is produced from 16 kW of electrical energy?

_____ 14. How much air is required to completely burn 5 cu ft of natural gas?

_____ 15. The amount of heat that is produced from .4 kW of electrical energy is _____ Btu.

_____ 16. How much heat is produced when 12 therms of natural gas is burned?

_____ 17. The amount of air required to completely burn 64 cu ft of propane is _____ cu ft.

_____ 18. If .5 gal. of Grade No. 2 fuel oil is burned at 100% efficiency, how much heat is produced? Use the standard heating value for Grade No. 2.

_____ 19. How much heat is produced when 90 lb of coal is burned at 100% efficiency?

_____ 20. How much heat is produced when 27 cu ft of butane is burned at 100% efficiency?

Matching

_____	**1.** Grade No. 1	**A.**	Inner mantle
_____	**2.** Grade No. 2	**B.**	2500 Btu/cu ft
_____	**3.** Grade No. 5	**C.**	Tip of flame
_____	**4.** Grade No. 6	**D.**	950 – 1000 Btu/cu ft
_____	**5.** Natural gas	**E.**	Preheating required
_____	**6.** Propane	**F.**	Used in pot burners
_____	**7.** Butane	**G.**	Outer mantle
_____	**8.** Dark blue	**H.**	Used in atomizing burners
_____	**9.** Light blue	**I.**	3200 Btu/cu ft
_____	**10.** Yellow	**J.**	Bunker C

Short Answer

1. Define and explain the importance of heating value.

2. Explain combustion efficiency.

3. Define fossil fuel.

4. Define ignition temperature.

5. Define complete combustion.

Name _Thomas O'Connell_   Date _1/31/2012_

True-False

(T) F   **1.** Ohm's law is the relationship between voltage, current, and resistance in an electrical circuit.

(T) F   **2.** In a parallel circuit, current may be flowing in one part of the circuit even though another part of the circuit is turned OFF.

~~T~~ (F)   **3.** A switch is a device that converts AC voltage to DC voltage by allowing the voltage and current to move in only one direction.

T (F)   **4.** A transformer is an electric device that uses direct current to change voltage from one level to another.

(T) F   **5.** In a series circuit, opening the circuit at any point stops the flow of current in all points of the circuit.

T (F)   **6.** The total current in a parallel circuit equals the sum of the current through all the loads.

T (F)   **7.** Control circuit transformers are designed to isolate the load from the power source.

(T) F   **8.** A selector switch is a switch with an operator that is rotated to activate the electrical contacts.

(T) F   **9.** In a parallel circuit, the voltage across each load remains the same if parallel loads are added or removed.

(T) F   **10.** An ON-delay timer delays for a predetermined time after receiving a signal to activate or turn ON.

Completion

Series   **1.** The sum of all the voltage drops in a(n) _____ circuit is equal to the applied voltage because the supply voltage is divided across all the loads of the circuit.

parallel   **2.** A(n) _____ circuit has two or more components connected so that there is more than one path for current flow.

polarity   **3.** _____ is the positive (+) or negative (–) electrical state of an object.

interrupt   **4.** Circuit breakers _____ circuit power when a predetermined value of current has been exceeded.

filtered   **5.** The output of a bridge rectifier is a pulsating DC voltage that must be _____ before it can be used in most electronic equipment.

relay **6.** A(n) _____ is a device that controls one electrical circuit by opening and closing contacts in another circuit.

AC **7.** _____ voltage reverses its direction of flow at regular intervals.

transformer **8.** A full-wave rectifier is a circuit containing two diodes and a center-tapped _____ that permits both halves of the input AC sine wave to pass.

push button **9.** A(n) _____ is a switch that opens or closes a circuit while manually pressed.

motor **10.** A(n) _____ is a machine that develops torque (rotating mechanical force) on a shaft, which is used to produce work.

Multiple Choice

A **1.** A(n) _____ circuit has two or more components connected so that there is only one path for current flow.
   A. series   C. series-parallel
   B. parallel   D. electronic

C **2.** The _____ across each load is the same when loads are connected in parallel.
   A. resistance   C. voltage
   B. current   D. energy

A **3.** The total resistance in a parallel circuit is _____ the smallest resistance value in the circuit.
   A. less than   C. equal to
   B. more than   D. none of the above

A **4.** _____ current flow is current flow from positive to negative.
   A. Direct   C. Electron
   B. Alternating   D. Conventional

D **5.** _____ energy is the energy of motion.
   A. Chemical   C. Static
   B. Potential   D. Kinetic

B **6.** _____ is the level of electrical energy in a circuit.
   A. Current   C. Resistance
   B. Voltage   D. Polarity

A **7.** A _____ is a switch that interrupts the supply of electric power from motors and machines.
   A. disconnect   C. full-wave rectifier
   B. half-wave rectifier   D. transformer

C **8.** A _____ switch detects the presence or absence of an object without touching the object.
   A. selector   C. proximity
   B. limit   D. temperature

B   **9.** A _____ is an electric output device that converts electrical energy into a linear mechanical force.

       A. rectifier   C. transformer

       B. solenoid   D. contactor

C   **10.** A bridge rectifier is a circuit containing _____ diodes that permits both halves of the input AC sine wave to pass.

       A. one   C. four

       B. two   D. eight

Calculations

 1. What is the total resistance of a series circuit consisting of a 12 Ω resistor, a 48 Ω solenoid valve, and a 100 Ω one-phase motor?

$$100 + 12 + 48 = 160$$

.4 Amps **2.** What is the total current flow in a series circuit that includes a 24 V power supply, a 48 Ω solenoid valve, and a 12 Ω resistor?

$$24/60 = .4 \qquad 48 + 12 = 60$$

.1137 **3.** What is the total resistance of a parallel circuit consisting of a 12 Ω resistor, a 48 Ω solenoid valve, and a 100 Ω one-phase motor?

$$1/12 + 1/48 + 1/100 = .0853 + .0208 + .01 = .1137$$

~~.2274~~ .5 **4.** What is the current flow through the solenoid valve in a parallel circuit consisting of a 24 V power supply, a 12 Ω resistor, a 48 Ω solenoid valve, and a 100 Ω one-phase motor?

$$.1137/24 = 8.74$$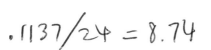

8.74 **5.** What is the total current flow in a parallel circuit consisting of a 24 V power supply, a 12 Ω resistor, a 48 Ω solenoid valve, and a 100 Ω one-phase motor?

Name _____   Date _____

Activity — Identification

Identify the type and describe the function of each switch.

(A)

(B)

Namco Controls Corporation
(C)

Banner Engineering Corp.
(D)

A. Name:

    Function:

B. Name:

    Function:

C. Name:

    Function:

D. Name:

    Function:

Activity — Measuring Voltage, Current, and Resistance

Follow all appropriate safety precautions when taking the following measurements with a digital multimeter.

Insert the test leads into an electrical outlet and measure the line voltage in a house, office, or classroom. Obtain a device with a heating element, such as a toaster, hair dryer, toaster oven, or electric heater. Unplug the device and measure the resistance of the device across the wires of the plug. The ON/OFF switch of the device will have to be ON for this to be a complete circuit.

_____   **1.** The measured voltage is _____ VAC.

_____   **2.** The measured resistance is _____ Ω.

Calculate the amount of current when the device when the device is operating.

_____   **3.** The calculated current is _____ A.

Plug in the device and turn it ON. Measure the current with a clamp-on ammeter attachment to the digital multimeter. Turn the device OFF and unplug it.

_____   **4.** The measured current is _____ A.

_____   **5.** Are the calculated current and measured current the same value?

Name _____ Date _____

True-False

T   F   **1.** A series circuit has two or more components connected so that there is more than one path for current flow.

T   F   **2.** Polarity is the positive (+) or negative (–) electrical state of an object.

T   F   **3.** A relay is a device that controls one electrical circuit by opening and closing contacts in another circuit.

T   F   **4.** The current through each load is the same when loads are connected in parallel.

T   F   **5.** Kinetic energy is the energy of motion.

Completion

_____   **1.** Ohm's law is the relationship between voltage, current, and _____ in an electrical circuit.

_____   **2.** In a(n) _____ circuit, current may be flowing in one part of the circuit even though another part of the circuit is OFF.

_____   **3.** A(n) _____ is a device that uses electromagnetism to change voltage from one level to another level.

_____   **4.** In a(n) _____ circuit, opening the circuit at any point stops the flow of current at all points in the circuit.

_____   **5.** A proximity switch detects the _____ or absence of an object without touching the object.

Multiple Choice

_____   **1.** A _____ is a device that converts AC voltage to DC voltage by allowing the voltage and current to move in only one direction.

        A. switch   C. filter

        B. rectifier   D. transformer

_____   **2.** _____ are designed to isolate the load from the power source.

        A. Parallel circuits   C. Control circuit transformers

        B. Solenoids   D. Isolation transformers

_____ **3.** A(n) _____ is a circuit containing two diodes and a center-tapped transformer that permits both halves of the input AC sine wave to pass.
   A. half-wave rectifier   C. bridge rectifier
   B. full-wave rectifier   D. isolation transformer

_____ **4.** What is the total resistance of a parallel circuit consisting of a 24 Ω resistor, a 48 Ω solenoid, and a 96 Ω pressure sensor?
   A. 168 Ω   C. 56 Ω
   B. 13.7 Ω   D. 24 Ω

_____ **5.** What is the total current flow in a parallel circuit consisting of a 24 V power supply, a 24 Ω resistor, a 48 Ω solenoid, and a 96 Ω pressure sensor?
   A. .14 A   C. 1.75 A
   B. .43 A   D. 24 A

Name _____   Date _____

Matching

_____   **1.** Atmospheric air   **A.** Moisture in air

_____   **2.** Dry air   **B.** Space occupied

_____   **3.** Moist air   **C.** Moisture begins to condense

_____   **4.** Properties of air   **D.** Actual moisture in air

_____   **5.** Temperature   **E.** Dry air and moisture

_____   **6.** Humidity   **F.** Characteristics

_____   **7.** Enthalpy   **G.** Air and moisture mixture

_____   **8.** Volume   **H.** Total heat content

_____   **9.** Dew point   **I.** Intensity of heat

_____   **10.** Humidity ratio   **J.** Air without moisture

True-False

T   F   **1.** A change in one property of air has no effect on the other properties.

T   F   **2.** A change in the dry bulb temperature of air does not change the relative humidity of the air.

T   F   **3.** A change in the wet bulb temperature of air indicates a change in the moisture content.

T   F   **4.** If two properties of air are known, the others can be found on a psychrometric chart.

T   F   **5.** Horizontal lines on a psychrometric chart indicate the dry bulb temperature of the air.

T   F   **6.** When air is cooled in an air conditioner, sensible and latent heat are involved.

T   F   **7.** When moisture is added to air by a humidifier, the dry bulb temperature does not change.

T   F   **8.** A psychrometric chart can be used to determine the amount of water that should be added to the air in a building to increase the relative humidity of the air.

T   F   **9.** The final condition of air during a mixing process cannot be determined using a psychrometric chart.

T   F   **10.** Any psychrometric chart may be used to find the properties of air at all atmospheric conditions and air pressures.

T   F   **11.** A psychrometer measures the enthalpy of the air.

T   F   **12.** The volume of air increases as air cools.

T   F   **13.** A change in wet bulb temperature changes the humidity ratio and relative humidity of the air.

T   F   **14.** Sensible heat cannot be measured with a thermometer or sensed by a person.

T   F   **15.** If dry bulb temperature changes, humidity ratio changes.

Completion

_____   **1.** Enthalpy is the sum of latent and _____ heat.

_____   **2.** _____ is the volume of a substance per unit of the substance.

_____   **3.** _____ air is the mixture of dry air, moisture, and particles.

_____   **4.** A(n) _____ is an instrument used for measuring humidity that consists of a wet bulb and a dry bulb thermometer mounted on a base.

_____   **5.** _____ is the amount of space occupied by a three-dimensional figure.

_____   **6.** _____ air does not contain moisture or particles.

_____   **7.** Wet bulb depression is the difference between wet bulb and _____ temperature readings.

_____   **8.** A(n) _____ is a chart that defines the condition of air at various properties.

_____   **9.** _____ are values used for comparing properties of air at different elevations and pressures.

_____   **10.** _____ heat is identified by a change of state and no temperature change.

_____   **11.** Air at a relative humidity of 46% holds _____% of the moisture it can hold at the same temperature if it were saturated.

_____   **12.** Enthalpy is expressed in _____.

_____   **13.** Relative humidity is _____% at the saturation line.

_____   **14.** Humidity ratio is expressed in grains per pound or _____ of moisture per pound of dry air.

_____   **15.** _____ heat is identified by the amount of water vapor in the air.

Psychrometrics

Name _____   Date _____

Activity — Dry Bulb Temperature/Humidity Ratio

Create a psychrometric chart by plotting dry bulb (db) temperature against humidity ratio (W) on the chart. Plot dry bulb temperature on the x-axis and humidity ratio on the y-axis of the chart. Enter dry bulb temperatures from 40°F on the left side to 110°F on the right side of the chart. Plot humidity ratio in grains of moisture per pound of dry air. Enter humidity ratio values on the right side of the chart from 0 gr/lb on the bottom to 200 gr/lb on the top of the chart.

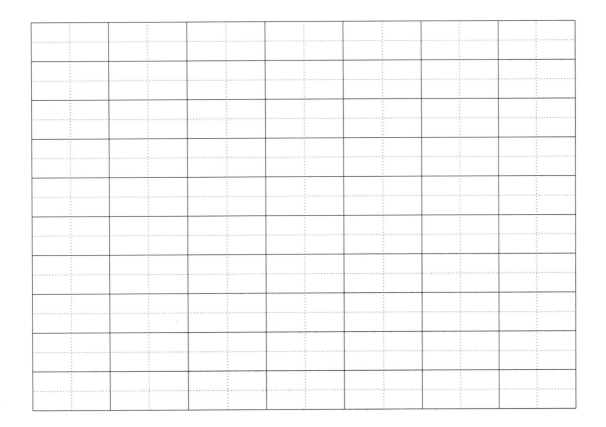

1. Find the point at the intersection of the 72°F db temperature line and the 63 gr/lb humidity ratio line. Mark the intersection as point 1.

2. Find the point at the intersection of the 96°F db temperature line and the 126 gr/lb humidity ratio line. Mark the intersection as point 2.

Activity — Saturation Line

At the following dry bulb temperatures, air is saturated with moisture at the specified humidity ratio. Using the dry bulb temperature and humidity ratio scales, plot the given points on the chart. Connect the points to create the curved saturation line from 40°F to 85°F. From the 85°F dry bulb temperature line, draw a horizontal line across the top of the chart.

Dry Bulb Temperature (in °F)	Humidity Ratio (in gr/lb)
40	36
50	54
60	78
70	111
80	156
85	185

_____ **1.** The humidity ratio at 46°F db and 100% relative humidity is _____ gr/lb.

_____ **2.** What is the db temperature of the air at saturation if the air holds 82 gr/lb of moisture?

Activity — Relative Humidity

Divide several db temperature lines from the bottom of the chart to the saturation line into ten equal segments. Connect the points between db temperature lines with curving lines from the bottom left of the chart to the top right. These lines indicate relative humidity (rh) in 10% increments.

_____ **1.** The relative humidity at 53°F db and 35 gr/lb is _____%.

_____ **2.** The humidity ratio at 80°F db and 50% rh is _____ gr/lb.

Activity — Wet Bulb Temperature

The wet bulb (wb) temperature lines begin on the curve of the chart and slope down to the right at a 23° angle. The scale for the wet bulb temperature lines is located on the curve of the chart. The wet bulb temperature and dry bulb temperature values are the same at the point where the lines intersect the saturation line. Draw the wet bulb temperature lines on the chart.

_____ **1.** What is the humidity ratio at a wet bulb temperature of 54°F db and 40% rh?

_____ **2.** The dry bulb temperature at 66°F wb and 30% rh is _____°F.

Activity — Enthalpy

The enthalpy (h) scale is located to the left of the curve of the chart. Enthalpy values are taken from lines extended from the wet bulb temperature lines. Draw a line parallel to the saturation line. Begin the enthalpy scale by using values of 15 Btu/lb at 39°F wb and 50 Btu/lb at 86°F. Divide the line into seven equal segments and enter the enthalpy values in 5 Btu/lb increments.

_____ **1.** The enthalpy at 80°F db and 70°F wb is _____ Btu/lb.

_____ **2.** Point 1 is found at 90°F db and 77°F wb. Point 2 is found at 60°F db and 50°F wb. Find the enthalpy difference between points 1 and 2 (heat removed).

Psychrometrics

Trade Test

Name _Thomas O'Connell_   Date _2/14/12_

Short Answer

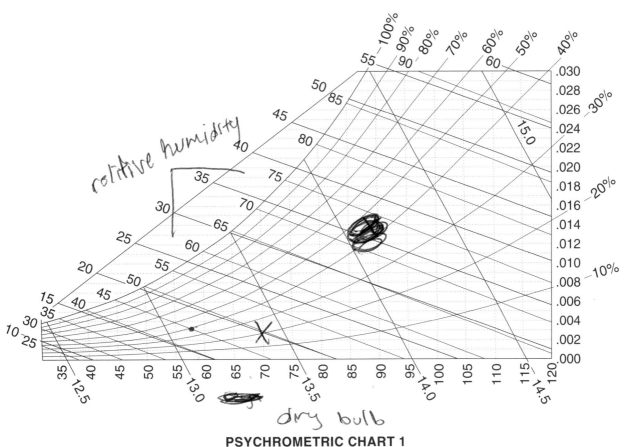

PSYCHROMETRIC CHART 1

Refer to Psychrometric Chart 1 on page 39.

Yes anything below dewpoint will condense

1. The air in a room is at a temperature of 73°F db and relative humidity of 46%. The glass in a window in the room is 40°F db. Will moisture condense on the inside of the window? Explain.

No, the humidity is too low relative to the room temp and window temp.

39

2. Place an X on the psychrometric chart on the point that defines standard air conditions.

66°

3. Two properties of the air are 67°F db and 63°F wb. The dew point temperature of the air is _____ °F.

85°

4. Air enters the heat exchanger of a furnace at 67°F and 45% relative humidity. The air is heated 17.5°F. The final dry bulb temperature is _____ °F.

5. A humidifier adds 17.5 gr/lb of moisture to the air. The air is at 57°F db and 40% rh. Find the following:

   57°
   A. db

   45°
   B. wb

   66%
   C. rh

   57°
   D. dp

   45
   E. h

   13°

6. Air enters an air conditioner cooling coil at 78°F db and 65°F wb. The air leaves the coil at 61°F db and 80% rh. The final wet bulb temperature is _____ °F.

14°

7. Air enters an air conditioner cooling coil at 84°F db and 70°F wb. The air leaves the coil at 70°F db and 55% rh. _____ Btu/lb of heat is removed from the air.

4960

8. An air conditioner removes 6.2 Btu/lb of air. The rate of cooling is 800 lb of air per minute. The cooling effect produced is _____ Btu/hr.

35

9. The air in a building is at 70°F db and 40% rh. The building is 50′ long, 40′ wide, and has a 9′ high ceiling. To raise the relative humidity in the building to 50%, _____ gal. of water is required.

10. Thirty percent makeup air at 55°F db and 50% rh is mixed with 70% return air at 85°F db and 65°F wb. Find the following:

   57°
   A. db

   45°
   B. wb

   40%
   C. rh

   57°
   D. dp

   45
   E. h

Name _____ Date _____

Multiple Choice

_____ 1. The cabinet of a furnace is used to _____.
A. keep dust out   C. A and B
B. separate flue gas and air   D. none of the above

_____ 2. A _____ centrifugal blower is furnished with a typical furnace.
A. backward-curved   C. forward-curved
B. vane-axial   D. propeller

_____ 3. The heat exchanger in a combustion furnace primarily _____.
A. separates flue gas and air   C. provides more heating surface
B. allows heat transfer   D. provides space for excess flue gas

_____ 4. A _____ changes line voltage to low voltage in control systems.
A. relay   C. transformer
B. thermostat   D. circuit breaker

_____ 5. A _____ is not a component of a furnace.
A. cabinet   C. distribution system
B. heat exchanger   D. filter

_____ 6. Area of influence is the area from the front of a register to a point where the air velocity drops below _____ fpm.
A. 50   C. 100
B. 75   D. 500

_____ 7. Most natural gas burners operate with a pressure of _____″ WC.
A. .1   C. 10.5
B. 3.5   D. 13.5

_____ 8. While operating, most LP gas burners have a pressure of _____″ WC.
A. .1   C. 10.5
B. 3.5   D. 15.5

_____ 9. A _____ igniter is a device that uses a small piece of silicon carbide that glows when electric current passes through it.
A. hot surface   C. glow plug
B. silicon   D. furnace

_____ 10. Thermostat _____ is the reduction in heating setpoint at night when occupants are asleep or the space is unoccupied.
A. control   C. adjustment
B. setback   D. tuning

Completion

_____ 1. Two blower drives used on furnaces are direct and _____ drive.

_____ 2. Two heat exchangers used in furnaces are clam shell and _____.

_____ 3. Two burners used with gas fuels are atmospheric and _____.

_____ 4. An electromagnetic _____ uses a magnetic coil to open or close one or more sets of contacts.

_____ 5. A(n) _____ burner is a small burner that is used to ignite the air-fuel mixture when the gas fuel valve opens.

_____ 6. _____ air distribution systems have a main supply duct and branch ducts.

_____ 7. A(n) _____ connects the supply air opening to the supply ductwork.

_____ 8. A(n) _____ allows dilution air to mix with the flue gas as the flue gas leaves the heat exchanger.

_____ 9. The _____ is the difference between the temperatures at which a furnace thermostat turns the burner(s) ON and OFF.

_____ 10. _____ pressure produces the pressure that forces combustion air into an atmospheric burner.

Short Answer

1. List the two ways that an atmospheric burner gets combustion air.

2. How does a power burner get combustion air?

3. What is the function of a pilot burner on a gas-fired furnace?

4. List the three major parts of a central forced-air heating system.

5. Explain why a furnace with a condensing heat exchanger requires a condensate drain.

6. List the four major parts of an air distribution system.

7. What are the advantages of a rooftop unit?

8. Explain setpoint temperature.

9. List four combustion safety controls.

10. How is a limit switch used in a furnace?

11. List four fuels used in combustion furnaces.

12. List four performance characteristics found on a blower performance chart.

13. List the four parts of an atmospheric burner.

14. How is the quantity of primary air adjusted on an atmospheric burner?

15. Explain how an electric spark igniter starts combustion in a burner.

True-False

T   F   **1.** A draft diverter eliminates downdrafts.

T   F   **2.** In a furnace, products of combustion and heated air are always completely separate.

T   F   **3.** Air distribution systems do not require filters.

T   F   **4.** Operating controls control the flow of electricity to a forced-air heating system.

T   F   **5.** A heat exchanger is the main component of a heating system, and the blower is the main component of a forced-air distribution system.

T   F   **6.** Fuses and circuit breakers protect against overcurrent.

T   F   **7.** Registers return air to the system blower.

T   F   **8.** Operating controls control the flow of electricity to a furnace.

T   F   **9.** A supply plenum connects the furnace supply air opening to the supply ductwork.

T   F   **10.** A direct-fired heater heats by radiation only.

T   F   **11.** Safety controls monitor the operation of a furnace.

T   F   **12.** A magnetic starter is a contactor that has overload relays added to it.

T   F   **13.** Combustion and electricity produce heat for furnaces.

T   F   **14.** Bimetal elements, remote bulbs, or electronic sensors can be used as a sensor in a thermostat.

T   F   **15.** A fan or blower in an atmospheric burner supplies and controls combustion air.

Matching

_____   **1.** Upflow furnace   **A.** Heater installed in a duct

_____   **2.** Infrared radiant heater   **B.** Fiberglass filter media

_____   **3.** Low-efficiency filter   **C.** Self-contained heating unit

_____   **4.** Horizontal furnace   **D.** Ductwork located below furnace

_____   **5.** Duct heater   **E.** Furnace installed in an attic

_____   **6.** Downflow furnace   **F.** Fibrous mat filter media

_____   **7.** Direct-fired heater   **G.** Bag-shape filter media

_____   **8.** High-efficiency filter   **H.** Heats by radiation only

_____   **9.** Unit heater   **I.** Ductwork located above furnace

_____   **10.** Medium-efficiency filter   **J.** Requires 100% makeup air

Name _____   Date _____

Activity — Identification

_____ **1.** Grill

_____ **2.** Register

_____ **3.** Return air ductwork

_____ **4.** Supply air ductwork

_____ **5.** Furnace

_____ **6.** Spud

_____ **7.** Port

_____ **8.** Manifold

_____ **9.** Burner tube

_____ **10.** Inlet vane

_____ **11.** Scroll

_____ **12.** Motor

_____ **13.** Cutoff plate

_____ **14.** Blower wheel

_____ **15.** Burner

_____ **16.** Controls

_____ **17.** Heat exchanger

_____ **18.** Blower

Carrier Corporation

Activity — Furnace Nameplate

Refer to the Furnace Nameplate on page 46.

AIRCO	UPFLOW FURNACE	
MODEL NO. 6AXW-20	SERIAL NO. 6362-89	
INPUT RATING	300,000	BTU/HR
OUTPUT RATING	231,000	BTU/HR
EFFICIENCY RATING	77	%
TYPE OF GAS	NAT	
ELECTRIC RATING	5.0	AMPS
120 V, 60 Hz, 1 PHASE		
MANIFOLD PRESS.	3.5	IN. WC

_____ 1. What is the input rating?

_____ 2. What is the output rating?

_____ 3. What is the efficiency rating?

_____ 4. What type of fuel is used?

_____ 5. What is the model number?

Activity — Furnace Ratings

Operate an installed furnace by turning the thermostat down to call for heat.

_____ 1. The temperature of the return air as it enters the furnace is _____ °F.

_____ 2. The temperature of the supply air as it leaves the furnace is _____ °F.

_____ 3. In a combustion furnace, what percentage of the heat produced by the combustion process leaves the furnace as warm air for heating a building?

4. What happens to the percentage of heat that does not heat a building?

5. Why does a high-efficiency combustion furnace have a higher efficiency rating than a conventional combustion furnace?

Activity — Blower Performance

Locate a furnace with a forward-curved centrifugal blower and a switch that turns the blower ON and OFF. Set the switch so the blower stays ON.

1. What causes air to be drawn into the return side of the blower?

2. What causes the air to be discharged from the blower?

Name _____   Date _____

Calculations

_____   **1.** An electric motor that turns at 1700 rpm has a 3.5″ motor sheave and is connected to a blower with a 7″ blower sheave. Find the blower wheel speed.

_____   **2.** A furnace with an input rating of 150,000 Btu/hr has an efficiency rating of 80%. The output rating of the furnace is _____ Btu/hr.

_____   **3.** Find the input rating for a fuel oil-fired furnace that burns Grade No. 2 fuel oil at a rate of 2 gph. Use the nominal heating value for Grade No. 2 fuel oil.

_____   **4.** Find the input rating for a gas-fired furnace that burns 80 cu ft of natural gas per hour. Use the nominal heating value for natural gas.

_____   **5.** Find the output rating of a combustion furnace that has an input rating of 100,000 Btu/hr and is 85% efficient.

_____   **6.** An electric motor turns at 1890 rpm, has a 4″ motor sheave, and is connected to a blower with a 6″ blower sheave. The speed of the blower wheel is _____ rpm.

_____   **7.** The blower in a furnace has a 2″ motor sheave and a 4″ blower sheave. The motor turns at 1200 rpm. The blower wheel speed is _____ rpm.

_____   **8.** A furnace has an input rating of 76,000 Btu/hr and output rating of 57,000 Btu/hr. Find the efficiency rating of the furnace.

_____   **9.** The input rating for a gas-fired furnace that burns propane at a rate of 60 cu ft/hr is _____ .

_____   **10.** A combustion furnace has an input rating of 120,000 Btu/hr and is 72% efficient. The output rating of the furnace is _____ Btu/hr.

_____   **11.** The furnace in a building has an input rating of 200,000 Btu/hr and output rating of 177,000 Btu/hr. Find the efficiency rating of the furnace.

_____   **12.** An industrial furnace has an input rating of 1,200,000 Btu/hr. The furnace is 78% efficient. The output rating of the furnace is _____ Btu/hr.

_____   **13.** The blower in a furnace has a 3″ motor sheave and a 7.5″ blower sheave. The motor turns at 3450 rpm. Find the blower wheel speed.

_____   **14.** A gas-fired furnace burns butane at a rate of 30 cu ft/hr. The input rating of the furnace is _____ Btu/hr.

_____   **15.** A combustion furnace has an input rating of 80,000 Btu/hr and is 94% efficient. The output rating of the furnace is _____ Btu/hr.

_____   **16.** The blower in a furnace has a 3.5″ sheave. The motor that turns the blower has a 2″ sheave. Find the speed of the motor if the blower wheel turns at 1100 rpm.

_____ 17. Find the efficiency of a furnace that has an input rating of 220,000 Btu/hr and an output rating of 176,000 Btu/hr.

_____ 18. The blower in a gas-fired furnace turns at 1600 rpm. The motor turns at 3200 rpm. Find the diameter of the motor sheave if the blower sheave is 5″.

_____ 19. A furnace burns natural gas at a rate of 150 cfm. Find the input rating of the furnace. Use the nominal heating value for natural gas.

_____ 20. The furnace in a building has an input rating of 95,000 Btu/hr and output rating of 78,000 Btu/hr. The efficiency rating of the furnace is _____%.

Identification

Identify the type of control by placing a P (power), O (operating), S (safety), or C (combustion safety) in front of the control.

_____ 1. Fuse

_____ 2. Flame surveillance

_____ 3. Thermostat

_____ 4. Circuit breaker

_____ 5. Blower control

_____ 6. Transformer

_____ 7. Relay

_____ 8. Limit switch

_____ 9. Stack switch

_____ 10. Disconnect

_____ 11. Contactor

_____ 12. Magnetic starter

_____ 13. Solenoid

_____ 14. Flame rod

Matching

_____ 1. Atmospheric burner

_____ 2. Combustion furnace

_____ 3. Modular heating system

_____ 4. Pulse burner

_____ 5. Electric furnace

_____ 6. Central heating system

_____ 7. Forced-air heating system

_____ 8. Power burner

_____ 9. Electrostatic filter

_____ 10. Prefilter

A. Heats one zone

B. Uses air to carry heat

C. Located before other filters

D. Chemical reaction

E. Heats entire building

F. Has no blower

G. Resistance heating elements

H. Highest filtering efficiency

I. Uses no excess air

J. Has combustion air blower

Name _____ Date _____

Completion

_____ 1. A condensate return system is a system used to return condensate back to the _____ after all the useful heat has been removed.

_____ 2. Boilers that produce heat by combustion are made of steel or _____.

_____ 3. The relationship between the water in a boiler and the firebox determines whether a boiler is a dry base boiler, a wet leg boiler, or a(n) _____ boiler.

_____ 4. A circulating pump is normally connected close to the boiler on the _____ piping.

_____ 5. The _____ of a boiler contains the hot products of combustion and the waterside contains water.

_____ 6. _____ are components directly attached to a boiler and boiler devices that are required for the operation of the boiler.

_____ 7. A flow _____ valve is a valve that regulates the flow of water in a hydronic heating system.

_____ 8. _____ heating is heat transfer that occurs when currents circulate between warm and cool regions of a fluid.

_____ 9. A(n) _____ valve may be completely open, completely closed, or at any intermediate position in response to the control signal the valve receives.

_____ 10. A(n) _____ steam heating boiler is a boiler operated at pressures not to exceed 15 psi of steam.

True-False

T   F   1. A hydronic heating system uses a boiler to store and add heat to water.

T   F   2. The pressure rating of a safety valve is the amount of steam that the safety valve is capable of venting at rated pressure.

T   F   3. A water column is a bent tube installed between the boiler and a boiler fitting to provide a water seal and prevent direct contact with steam.

T   F   4. A gauge glass is a tubular glass column that indicates the water level in the boiler.

T   F   5. A steam header is a distribution pipe that supplies steam to branch lines.

T   F   6. An air vent allows the water in a hydronic heating system to expand without raising the system water pressure to dangerous levels.

T   F   **7.** A four-pipe piping system combines the return pipe for the heating and cooling systems.

T   F   **8.** A terminal device is a heating system component that transfers heat from the hot water in a hydronic heating system to the air in building spaces.

T   F   **9.** Deadband is the thermostat setpoint differential when no heating or cooling is allowed because the temperature is between the two setpoints.

T   F   **10.** A pressure-temperature gauge is a gauge that measures the pressure and temperature of the water at the point on the boiler where the gauge is located.

Multiple Choice

_____ **1.** A _____ boiler heats water that surrounds the tubes as the hot gases of combustion pass through the fire tubes.

    A. firetube   C. cast iron
    B. watertube   D. steel

_____ **2.** A(n) _____ gauge is a pressure gauge that indicates vacuum in inches of mercury and pressure in psi on the same gauge.

    A. steam pressure   C. pressure-temperature
    B. compound   D. operating

_____ **3.** A _____ is a warning device used to indicate extreme overheating of the boiler from a low water condition.

    A. fusible plug   C. bottom blowdown valve
    B. vent valve   D. fitting

_____ **4.** A _____ valve controls the flow of makeup water into a boiler to compensate for losses.

    A. check   C. safety
    B. gate   D. feedwater

_____ **5.** A fuel oil _____ is a boiler component that provides atomized fuel oil to the boiler.

    A. strainer   C. pump
    B. gas valve   D. burner

_____ **6.** _____ draft is a mechanical draft created by air pulled through the boiler firebox by a blower located in the breaching after the boiler.

    A. Natural   C. Induced
    B. Forced   D. Balanced

Short Answer

1. Should an expansion tank in a hydronic heating system be filled with water? Explain.

2. What is the difference between a firetube boiler and a watertube boiler?

3. Define hydronic heating system.

4. What is the difference between a globe valve and a gate valve?

5. How can try cocks be used to determine the level of water in a boiler?

Name _____   Date _____

Activity — Identification

_____ **1.** Circulating pump

_____ **2.** Fitting

_____ **3.** Controls

_____ **4.** Piping

_____ **5.** Boiler

_____ **6.** Terminal device

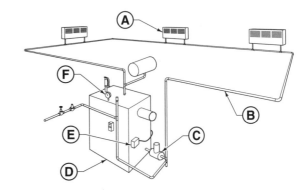

_____ **7.** Unit ventilator

_____ **8.** Baseboard convector

_____ **9.** Unit heater

_____ **10.** Cabinet convector

_____ **11.** Cabinet heater

_____ **12.** Radiant panel

_____ **13.** Surface radiation

_____ **14.** Standing radiator

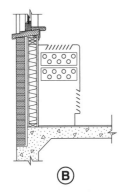

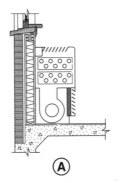

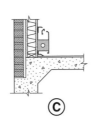

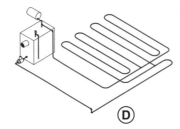

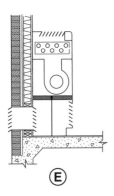

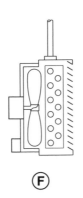

Activity — Valve Function

Briefly describe each valve and its function in a hydronic piping system.

1. Blowdown

2. Check

3. Pressure-reducing

4. Spring safety

5. Flow control

Activity — Piping Systems

Draw the piping for a two-pipe direct-return hydronic distribution system by connecting the parts with lines. On each section of pipe, mark the direction of water flow with arrows. Briefly describe the disadvantage of the piping system.

1. Disadvantage:

SUPPLY PIPING ————————

RETURN PIPING - - - - - - - - - - - - - - - -

Name _____   Date _____

Short Answer

1. If a boiler is shut OFF at night, what pickup factor is required to warm the water in the hydronic heating system?

2. How is air removed from the water in a hydronic system?

3. Which piping system is the most economical to operate if both heating and cooling are required in different parts of a building simultaneously? Why?

4. Which terminal device is most appropriate for spot heating applications? Why?

5. Which terminal device is installed when ventilation air is required? Why?

6. What is the minimum number of safety valves required for a boiler by the ASME Code?

7. List four heating applications for low-pressure steam boilers.

8. How is the output rating of a cast iron (sectional) boiler increased or decreased?

9. Why is makeup water added to a boiler?

10. Why is the output rating of electric boilers 100%?

11. Why are boiler and terminal device controls required on hydronic heating systems?

12. List the four ways boilers are classified.

13. Describe the function of a zone valve.

14. Describe a typical circulating pump used in a hydronic piping system.

15. Describe the difference between a one-pipe series and one-pipe primary-secondary piping system.

Name _Thomas O'Connell_   Date _____

Short Answer

1. Why should a refrigerant have a low freezing point?

   So they can boil at room temp.

2. Why should a refrigerant have a high latent heat?

   because It is changing from liquid to vapor.

3. What is a refrigerant property table? What type of information does a refrigerant property table contain?

   pressures and boiling points for various refrigerants

4. What is the basic principle on which an absorption refrigeration system produces a refrigeration effect?

   It absorbs heat transfering energy from hot to cold.

5. Define superheat and subcooling.

   superheat is sensible heat after substance has turn to vapor.
   Subcooling is cooling to temp lower than saturation temp.

Completion

refrigerant 1. A(n) _____ is the medium that changes state in an operating refrigeration system.

compressor 2. A(n) _____ is the component in a mechanical compression refrigeration system that raises the pressure and temperature of the refrigerant.

metering device 3. In a mechanical compression refrigeration system, a(n) _____ causes a pressure decrease in the refrigerant.

evaporator 4. A(n) _____ is the component in a refrigeration system that removes heat from air or water.

saturated 5. A refrigerant stored in a cylinder is in the ___saturated___ state.

decreases 6. If the pressure on a liquid decreases, the boiling point temperature _____.

increases 7. If the pressure on a liquid increases, the boiling point temperature _____.

chlorine 8. The two main chemical components of refrigerants used in modern refrigeration systems are _____ and fluorine.

atmospheric 9. In a mercury barometer, _____ pressure holds the mercury up in the tube.

methane 10. A refrigerant that has a two-digit refrigerant number is derived from _____.

water 11. A chiller is a piece of refrigeration equipment that removes heat from _____ that circulates through a building for cooling purposes.

centrifical 12. Low-pressure chillers use _____ compressors.

condenser 13. A(n) _____ is a component in a chiller that transfers the heat from the refrigerant to a cooling medium, normally water but sometimes air.

evaporitor 14. A(n) _____ is the component in the chiller system that transfers heat from the water to the liquid refrigerant.

air pressure 15. A purge unit is a device used to maintain a system free of _____ and moisture.

Multiple Choice

B 1. When a liquid evaporates, the temperature of the remaining liquid _____.
   A. increases   C. decreases
   B. remains constant   D. increases or remains constant

C 2. Heat flows from _____ to _____ in the condensing section of an operating refrigeration system.
   A. air; water   C. refrigerant; air
   B. air; oil   D. air; refrigerant

B 3. The second digit in a two-digit refrigerant number represents the number of _____ atoms in each molecule of the refrigerant.
   A. hydrogen   C. oxygen
   B. fluorine   D. carbon

D 4. The two components that maintain a pressure difference in a refrigeration system are the _____ and _____.
   A. evaporator; compressor   C. evaporator; condenser
   B. expansion device; condenser D. expansion device; compressor

A 5. A condenser _____ refrigerant in a refrigeration system.
   A. rejects heat from   C. adds heat to
   B. compresses   D. none of the above

True-False

T (F) 1. Refrigerant numbers are assigned at random and have no meaning.

(T) F 2. Fluorine and chlorine atoms replace the hydrogen atoms in methane or ethane to produce halocarbon refrigerants.

(T) F 3. In the condenser of an absorption refrigeration system, heat is added to the refrigerant-absorbent solution to vaporize the refrigerant and raise its temperature and pressure.

(T) F 4. An absorption refrigeration system produces a refrigeration effect with mechanical equipment.

T (F) 5. According to the second law of thermodynamics, heat always flows from a material at a low temperature to a material at a high temperature.

9 Refrigeration Principles

Name _____   Date _____

Activity — Sensible Heat

Partially fill a container with water. Measure and record the temperature of the water.

_____   **1.** The temperature of the water is _____°F.

Place the container over a heat source. Observe the temperature increase of the water as heat is added. After the water begins to boil, notice that the temperature does not increase.

2. What change is taking place in the water?

3. In what part(s) of a refrigeration system does a similar change occur?

Activity — High- and Low-Pressure Sides

Identify the high- and low-pressure sides of the mechanical compression refrigeration system. Draw a line across the diagram to separate the sides and label the two sides.

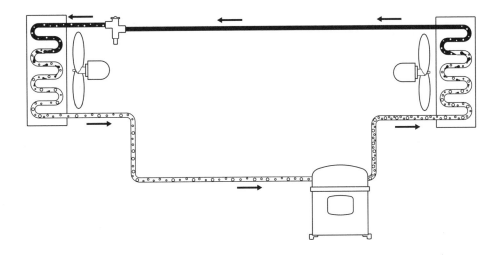

MECHANICAL COMPRESSION REFRIGERATION SYSTEM

Activity — Pressure-Temperature Relationships

Make sure the valves on a manifold set are closed. Attach the hose on the high-pressure side of a manifold set to the valve on a refrigerant cylinder. Use a cylinder of refrigerant at room temperature. Open the valve on the cylinder slowly. Record the pressure on the manifold set gauge and the temperature of the room.

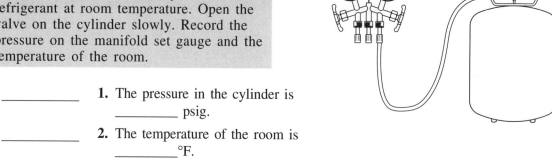

_____ 1. The pressure in the cylinder is _____ psig.

_____ 2. The temperature of the room is _____ °F.

Observe the pressure-temperature relationship on a refrigerant property table for the refrigerant in the cylinder.

3. At the recorded pressure, is the temperature shown on the table the same as room temperature?

4. If the pressure in the cylinder is reduced by opening the valve, what happens to the temperature of the refrigerant?

5. What happens to the pressure of the refrigerant over a period of time as the temperature of the refrigerant returns to room temperature?

Activity — Condensing Point Temperature

Partially fill a container with water. Record the temperature of the water.

_____ 1. The temperature of the water is _____ °F.

Fill the container with ice. Observe the outside of the container for moisture. Record the temperature of the water when moisture appears.

_____ 2. The temperature of the water when moisture appears is _____ °F.

3. Where does the moisture on the outside of the container come from?

4. How is heat involved?

5. In what part(s) of a refrigeration system does a similar change occur?

Activity — System Operation

Attach a manifold set to a refrigeration or air conditioner trainer. Place thermometers at the air inlet and outlet of the evaporator and condenser coils. Turn the trainer ON and operate it long enough to stabilize the operation. Record the refrigerant pressures in the low- and high-pressure sides of the system.

_____ **1.** The low-pressure side pressure is _____ psig.

_____ **2.** The high-pressure side pressure is _____ psig.

Record the temperature of the refrigerant entering the evaporator and condenser coils.

_____ **3.** The temperature of the refrigerant entering the evaporator coil is _____°F.

_____ **4.** The temperature of the refrigerant entering the condenser coil is _____°F.

Record the temperature of the air entering and leaving the evaporator and condenser coils.

_____ **5.** The temperature of the air entering the evaporator coil is _____°F.

_____ **6.** The temperature of the air leaving the evaporator coil is _____°F.

_____ **7.** The temperature of the air entering the condenser coil is _____°F.

_____ **8.** The temperature of the air leaving the condenser coil is _____°F.

9. What is the relationship between the pressure and temperature of the refrigerant in the evaporator coil?

10. What is the relationship between the pressure and the temperature of the refrigerant in the condenser coil?

11. How is the temperature decrease of the air through the evaporator coil related to heat transfer?

12. How is the temperature increase of the air through the condenser coil related to heat transfer?

Activity — Evaporation Heat Loss

Obtain two shallow saucers. One saucer should be slightly smaller than the other. Partially fill the smaller saucer with water and place it in the larger saucer. Make sure the water does not rise above the top edge of the small saucer. Place or suspend a thermometer so that the bulb is in the water in the small saucer. Pour some aqueous ammonia, alcohol, or ether into the large saucer. Record the temperature of the water in the small saucer at 1 minute intervals for 10 minutes.

1. What happens to the temperature of the water?

2. What causes the change?

3. What is the boiling point temperature of the substance in the large saucer?

Time (in minutes)	Temperature (in °F)
Initial	
1	
2	
3	
4	
5	
6	
7	
8	
9	
10	

4. If the liquid in the large saucer had a lower boiling point, how would the time/temperature relationship change?

Name _____   Date _____

Calculations

_____ 1. The absolute pressure for a gauge pressure of 14.7 psig is _____ psia.

_____ 2. The refrigeration effect of a refrigerant that enters an evaporator of an absorption refrigeration system with an enthalpy of 75.9 Btu/lb and leaves with an enthalpy of 1060 Btu/lb is _____ Btu/lb.

_____ 3. A refrigeration system produces 61,000 Btu/hr of cooling while using the equivalent of 85,000 Btu/hr of energy. What is the COP of the system?

_____ 4. The absolute pressure for a gauge pressure of 0 psig is _____ psia.

_____ 5. A refrigerant enters an evaporator of an absorption refrigeration system with an enthalpy of 92 Btu/lb and leaves with an enthalpy of 1127 Btu/lb. The refrigeration effect of the evaporator is _____ Btu/lb.

_____ 6. What is the absolute pressure for a gauge pressure of 85.3 psig?

_____ 7. Refrigerant enters the condenser of an absorption system with an enthalpy of 1092 Btu/lb and leaves with an enthalpy of 80.9 Btu/lb. What is the heat of rejection?

_____ 8. A refrigeration system produces 170,000 Btu/hr of cooling while using the equivalent of 230,000 Btu/hr of energy. The COP of the system is _____.

_____ 9. A refrigerant enters the condenser of an absorption refrigeration system with an enthalpy of 902 Btu/lb and leaves with an enthalpy of 77 Btu/lb. The heat of rejection of the condenser is _____ Btu/lb.

_____ 10. A refrigeration system produces 24,000 Btu/hr of cooling while using the equivalent of 30,000 Btu/hr of energy. The COP of the system is _____.

Identification

1. Identify and describe the function of the components of a compression refrigeration system.

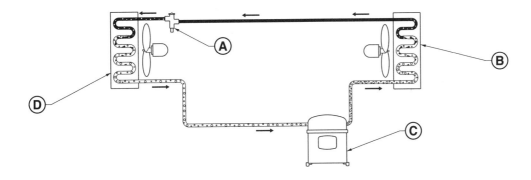

A. Name
   Function

B. Name
   Function

C. Name
   Function

D. Name
   Function

2. Identify and describe the main sections of an absorption refrigeration system.

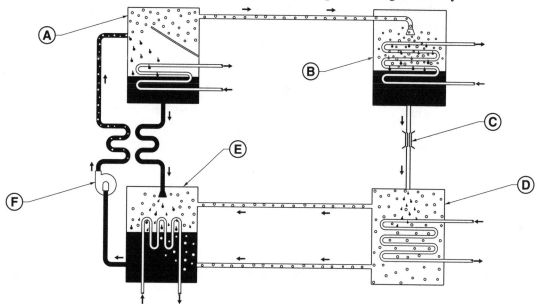

A. Name
   Function

B. Name
   Function

C. Name
   Function

D. Name
   Function

E. Name
   Function

F. Name
   Function

Name _____ Date _____

True-False

| T | F | **1.** Refrigerant management laws and regulations do not apply to motor vehicles. |

T   F   **1.** Refrigerant management laws and regulations do not apply to motor vehicles.

T   F   **2.** A gas cylinder must have a protective cap over the valve while the cylinder is being moved.

T   F   **3.** Disposable gas cylinders can safely be refilled with recovered refrigerant.

T   F   **4.** Fluorocarbon-based refrigerants can break down in the presence of an open flame and generate poisonous phosgene gas.

T   F   **5.** Type II certification is for technicians who service small appliances.

T   F   **6.** Recovered refrigerant must not be reused in any equipment other than equipment having the same owner as the equipment from which the refrigerant was removed.

T   F   **7.** Newer refrigerants are nontoxic and nonflammable, but they can still cause harm.

T   F   **8.** When adding or removing refrigerant, low-loss fittings are used to minimize the release of refrigerant to the atmosphere.

T   F   **9.** Lockout/tagout regulations are optional when a trained technician is working on refrigeration and air conditioning equipment.

T   F   **10.** When charging a system with refrigerant on the low side of the system, always charge as a liquid.

Completion

_____   **1.** _____ are procedures created by the EPA that require technicians that service and dispose of refrigeration and air conditioning equipment to observe specified procedures that reduce refrigerant emissions.

_____   **2.** When refrigerants are released to the atmosphere they can cause _____ to unprotected skin.

_____   **3.** Federal regulations require oxygen monitors and _____ to be installed in all spaces with refrigeration equipment or worn by HVAC technicians to warn of a refrigerant leak.

_____   **4.** Technicians must not vent, release, or dispose of any substance used as a refrigerant in a manner which permits such a substance to enter the _____.

_____   **5.** The EPA requires the evacuation of refrigeration and air conditioning equipment to established levels depending on the _____ of the evacuation equipment used.

Multiple Choice

_____   **1.** The EPA requires that all substantial leaks in systems with more than _____ lb of refrigerant charge must be repaired as soon as possible.

    A. 1   C. 50

    B. 5   D. 500

_____   **2.** Before welding, brazing, or soldering on a refrigeration system, the refrigerant must first be _____ and the system opened to the atmosphere.

    A. recovered   C. pressurized

    B. recycled   D. warmed

_____   **3.** When filling a recovery cylinder with liquid refrigerant, do not exceed _____ % of the cylinder's volume (by weight).

    A. 50   C. 100

    B. 80   D. 125

_____   **4.** When pressurizing a refrigeration system, use an inert gas such as _____ or carbon dioxide.

    A. air   C. oxygen

    B. propane   D. nitrogen

_____   **5.** _____ certification is for technicians that service high-pressure appliances.

    A. Type I   C. Type III

    B. Type II   D. Universal

_____   **6.** Owners of large refrigeration and air conditioning equipment are required to keep records of the amount of _____ added to equipment during maintenance and repair.

    A. refrigerant   C. hydraulic fluid

    B. oil   D. air

_____   **7.** Refrigerant _____ is when refrigerant is processed to new product specifications by means that may include distillation.

    A. removal   C. reclaiming

    B. recycling   D. recovery

10 Management of Refrigerants and Refrigeration Systems

Name _____ Date _____

Activity — Refrigerant Handling

After obtaining permission from the proper authority, inspect an air conditioning or refrigeration device used at work, in a school, or in a large office building. Complete the inspection report.

Date _____ Site _____

Person in charge _____ Permission granted by _____

Equipment manufacturer _____

1. What refrigerant is used in that device?

2. What type of certification, Class I, II, or III, is required to service or maintain that device?

3. What is the allowable annual leak rate for that device?

Name _____ Date _____

True-False

T   F   **1.** Technicians must not vent, release, or dispose of any substance used as a refrigerant in a manner which permits such a substance to enter the environment.

T   F   **2.** Federal regulations require oxygen monitors and alarm systems to be installed in all spaces with refrigeration equipment or worn by HVAC technicians to warn of a refrigerant leak.

T   F   **3.** When pressurizing a refrigeration system, use an inert gas such as acetylene.

T   F   **4.** Owners of large refrigeration and air conditioning equipment are required to keep records of the amount of hydraulic oil added to equipment during maintenance and repair.

T   F   **5.** Service practices are procedures created by the EPA that require technicians that service and dispose of refrigeration and air conditioning equipment to observe specified procedures that reduce refrigerant emissions.

Completion

_____   **1.** A gas cylinder must have a(n) _____ over the valve while the cylinder is being moved.

_____   **2.** When adding or removing refrigerant, _____ are used to minimize the release of refrigerant to the atmosphere.

_____   **3.** When charging a system with refrigerant on the low side of the system, always charge as a(n) _____.

_____   **4.** The EPA requires that all substantial leaks in systems with more than _____ lb of refrigerant charge must be repaired as soon as possible.

_____   **5.** Before welding, brazing, or soldering on a refrigeration system, the refrigerant must first be _____ and the system opened to the atmosphere.

Short Answer

1. List the four types of technician certification and explain the differences between them.

2. List the topics covered in certification tests.

3. Explain the procedures for safely handling refrigerant cylinders as they are being transported.

4. Explain the difference between the terms "refrigerant recovery," "refrigerant recycling," and "refrigerant reclaiming."

Name _____   Date _____

True-False

T   F   **1.** The refrigerant flowing through the evaporator of a refrigeration system is represented by a vertical line on a pressure-enthalpy diagram.

T   F   **2.** On a pressure-enthalpy diagram, a graph of the operation of a refrigeration system represents the properties of the refrigerant at all points in the system.

T   F   **3.** High compression ratio indicates an efficiently operating refrigeration system.

T   F   **4.** The same pressure-enthalpy diagram can be used for all refrigerants.

T   F   **5.** An ideal refrigeration cycle drawn on a pressure-enthalpy diagram shows superheat and subcooling of the refrigerant in the system.

T   F   **6.** The total heat rejected by a refrigeration system must equal the heat absorbed in the evaporator plus the heat of compression.

T   F   **7.** Seasonal energy efficiency is a performance rating of a refrigeration system that operates under normal conditions over a period of time.

T   F   **8.** Superheat is sensible heat added to a substance after it has changed state.

T   F   **9.** Heat of compression is the thermal energy equivalent of the mechanical energy used to compress the refrigerant.

T   F   **10.** Liquidation pressure is the pressure at which a substance changes state.

Short Answer

1. What is a pressure-enthalpy diagram?

2. Define compression ratio.

3. Explain the difference between an ideal and actual refrigeration cycle.

4. Explain constant volume.

5. Define saturation temperature.

Completion

_____ **1.** Specific volume is expressed in _____ on a pressure-enthalpy diagram.

_____ **2.** Constant _____ is a percentage that expresses the ratio of vapor to liquid as a refrigerant changes state.

_____ **3.** A(n) _____ tube is a tube inside a mechanical pressure gauge.

_____ **4.** _____ is the total heat contained in a substance.

_____ **5.** _____ pressure is the pressure of the refrigerant in a refrigeration system.

Multiple Choice

_____ **1.** _____ is the process of cooling a substance to a temperature that is lower than the saturation temperature of the substance at a given pressure.
   A. Vaporization
   B. Subcooling
   C. Superheating
   D. Compression

_____ **2.** Constant _____ is a calculated value that indicates energy lost to the disorganization of the molecular structure of a substance when heat is transferred.
   A. entropy
   B. quality
   C. volume
   D. enthalpy

_____ **3.** Heat of rejection is found by subtracting the enthalpy of the refrigerant as it enters the condenser from the enthalpy of the refrigerant as it leaves the _____.
   A. evaporator
   B. expansion device
   C. compressor
   D. condenser

_____ **4.** Refrigeration effect is the amount of heat absorbed by a refrigerant in the _____ of a refrigeration system.
   A. evaporator
   B. expansion device
   C. compressor
   D. condenser

_____ **5.** _____ is the cooling performance rating of a refrigeration system that operates under normal conditions over a period of time.
   A. COP
   B. SEER
   C. Compression ratio
   D. Refrigeration effect

Name _____   Date _____

Activity — Refrigerant Pressure

Make sure the valves on a manifold set are closed. Observe the pressure when the manifold set is disconnected. Record the pressure.

_____ **1.** The pressure with the manifold set disconnected is _____ psig.

Attach one end of a refrigerant hose to the manifold set and the other end to a refrigerant cylinder. Carefully open the valve on the refrigerant cylinder to allow the pressure in the cylinder to register on the gauge of the manifold set. Record the pressure.

_____ **2.** The pressure with the manifold set connected and refrigerant valve open is _____ psig.

3. What pressure is the gauge showing when it is not connected to a system? Explain.

4. Is 14.7 psi added to or subtracted from the gauge reading to change gauge pressure to absolute pressure?

Activity — Pressure/Temperature Relationships

Make sure the valves of a manifold set are closed. Connect the manifold set to a refrigerant cylinder at room temperature. Slowly open the valve on the refrigerant cylinder. Record the temperature of the air and pressure of the refrigerant.

_____ **1.** The air temperature is _____°F.

_____ **2.** The refrigerant pressure at room temperature is _____ psig.

Compare the recorded refrigerant pressure with a pressure-enthalpy diagram for the same refrigerant at the same temperature.

_____ **3.** The pressure on a pressure-enthalpy diagram is _____ psig at the same temperature.

Place the cylinder of refrigerant in a container of ice. After several minutes, record the pressure of the refrigerant.

_____ **4.** The pressure of the refrigerant is _____ psig.

5. Does the recorded pressure and temperature of the refrigerant in the cylinder at room temperature match that found on a pressure-enthalpy diagram?

6. What happens to the pressure in the cylinder when it is cooled?

Activity — Superheat

Partially fill a Pyrex® flask with water. Place the flask over a Bunsen burner. Insert a rubber stopper with two holes in the opening of the flask. Place a thermometer in one hole so that it measures the temperature of the water. Place a glass tube with a 90° bend in the other hole. Make sure the opening of the glass tube inside the flask is above the surface of the water in the flask. With this arrangement, part of the tube is horizontal. Place a second Bunsen burner under the horizontal section of tube. Suspend a second thermometer so that the bulb lies in the steam path and measures the temperature of the steam leaving the tube.

  Light the Bunsen burner under the flask. When the water begins to boil, record the temperature of the water in the flask and the temperature of the steam coming out of the tube. **Caution:** Care should be taken to ensure that steam can readily escape from the flask through the glass tube.

_____ **1.** The water temperature with the second Bunsen burner OFF is _____ °F.

_____ **2.** The steam temperature with the second Bunsen burner OFF is _____ °F.

Light the second Bunsen burner. After the tube becomes heated, record the temperature of the water and steam.

_____ **3.** The water temperature with the second Bunsen burner ON is _____ °F.

_____ **4.** The steam temperature with the second Bunsen burner ON is _____ °F.

5. With the second Bunsen burner OFF, is the temperature of the steam leaving the horizontal tube approximately the same temperature as the boiling water? Why or why not?

6. With the second Bunsen burner ON, what happens to the temperature of the steam leaving the horizontal tube compared to the temperature of the water? Why?

Activity — Subcooling

Fill a flexible container with water. Place a thermometer that reads temperatures below 32°F in the water. Place the container in a freezer. Set another thermometer in the same compartment to measure the air temperature in the freezer. Set the temperature dial on the unit below 32°F. When ice begins to form on the water, record the temperature of the water and the air.

_____ **1.** The water temperature as ice begins to form is _____°F.

_____ **2.** The air temperature as ice begins to form is _____°F.

Continue to check the temperatures as the water freezes solid. Just after all the water is frozen solid, record the water and air temperatures.

_____ **3.** The water temperature just after the water freezes is _____°F.

_____ **4.** The air temperature just after the water freezes is _____°F.

Turn the temperature dial on the unit to its lowest setting. Observe the temperature of the ice and air over several hours. Record the temperatures at 1 hour intervals.

5. Does the temperature of the water change as it freezes? Explain.

_____ **6.** The temperature of the air after 1 hour is _____°F.

_____ **7.** The temperature of the air after 2 hours is _____°F.

_____ **8.** The temperature of the ice after 1 hour is _____°F.

_____ **9.** The temperature of the ice after 2 hours is _____°F.

10. At what temperature does the temperature of the ice vary from the temperature of the air?

Activity — System Pressure

Attach a manifold set to a refrigeration or air conditioning trainer and turn the equipment ON. After the operation has stabilized, record the high- and low-pressure side pressures and temperatures.

_____ 1. The high-pressure side pressure is _____ psig.

_____ 2. The low-pressure side pressure is _____ psig.

_____ 3. The temperature of the refrigerant leaving the condenser coil is _____°F.

_____ 4. The temperature of the refrigerant leaving the evaporator coil is _____°F.

Plot the low-pressure line on a pressure-enthalpy diagram for the refrigerant. Plot the high-pressure line on a pressure-enthalpy diagram for the refrigerant. Connect the left ends of the low- and high-pressure lines with a vertical line. Connect the right ends of the low- and high-pressure lines with a line that runs parallel to the nearest constant entropy line. Refer to the appropriate pressure-enthalpy diagram on pages 173–187.

5. Circle the part of the high- and low-pressure lines that indicate the refrigerant in a saturated condition.

_____ 6. The amount of heat rejected by the refrigerant in the evaporator is _____ Btu/lb.

_____ 7. The amount of heat rejected by the refrigerant in the condenser is _____ Btu/lb.

_____ 8. The amount of heat added to the refrigerant in the compressor is _____ Btu/lb.

_____ 9. The amount of heat rejected by or added to the refrigerant in the expansion device is _____ Btu/lb.

_____ 10. What is the amount of heat rejected from the refrigerant in the condenser equal to?

_____ 11. The cooling capacity of the system is _____ Btu/hr.

_____ 12. The amount of superheat in the refrigerant leaving the evaporator is _____ Btu/lb.

_____ 13. The amount of subcooling in the refrigerant leaving the condenser is _____ Btu/lb.

_____ 14. The compression ratio of the system is _____.

_____ 15. The COP for the system is _____.

Name _____   Date _____

Calculations

Refer to the appropriate Pressure-Enthalpy Diagram on pages 173–187.

_____ 1. What is the absolute pressure when the low-pressure side of a refrigeration system is 76 psig?

_____ 2. The saturated pressure for R-502 at a temperature of 55°F is _____ psig.

_____ 3. What is the temperature of a refrigerant if 15°F of superheat is added at a saturated vapor temperature of 73°F?

_____ 4. The enthalpy of the refrigerant that enters the evaporator in a mechanical compression refrigeration system is 42 Btu/lb. The enthalpy of the refrigerant that leaves the evaporator is 146 Btu/lb. The refrigeration effect of the system is _____ Btu/lb.

_____ 5. A refrigeration system has a high-pressure side of 171 psig and a low-pressure side of 90 psig. The compression ratio is _____.

_____ 6. The refrigeration effect of a refrigeration system is 77 Btu/lb and the heat of compression is 13 Btu/lb. The COP of the refrigeration system is _____.

_____ 7. The enthalpy of the refrigerant that leaves the compressor in a mechanical compression refrigeration system is 157 Btu/lb. The enthalpy of the refrigerant that enters the compressor is 146 Btu/lb. The heat of compression is _____ Btu/lb.

_____ 8. The refrigeration effect of a refrigeration system is 100 Btu/lb and the heat of compression is 16 Btu/lb. The COP of the refrigeration system is _____.

_____ 9. The enthalpy of the refrigerant that leaves the compressor in a mechanical compression refrigeration system is 172 Btu/lb. The enthalpy of the refrigerant that enters the compressor is 139 Btu/lb. The heat of compression is _____ Btu/lb.

_____ 10. A refrigeration system has a high-pressure side of 155 psig and a low-pressure side of 82 psig. The compression ratio is _____.

_____ 11. The enthalpy of the refrigerant that enters the condenser in a mechanical compression refrigeration system is 167 Btu/lb. The enthalpy of the refrigerant that leaves the condenser is 55 Btu/lb. The heat of rejection is _____ Btu/lb.

_____ 12. The enthalpy of the refrigerant that enters the evaporator in a mechanical compression refrigeration system is 36 Btu/lb. The enthalpy of the refrigerant that leaves the evaporator is 150 Btu/lb. The refrigeration effect of the system is _____ Btu/lb.

Matching

_____ 1. Saturation temperature

_____ 2. Saturated liquid

_____ 3. Saturated vapor

_____ 4. Superheat

_____ 5. Refrigeration effect

A. Heat absorbed by refrigerant

B. Heat added after change of state

C. Temperature at change of state

D. Condensation occurs if temperature decreases

E. Vaporization occurs if temperature increases

Identification

_____ 1. Constant volume

_____ 2. Constant entropy

_____ 3. Enthalpy

_____ 4. Temperature

_____ 5. Pressure

_____ 6. Constant quality

Ⓐ

Ⓑ

Ⓒ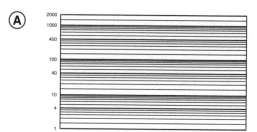

Ⓓ

Ⓔ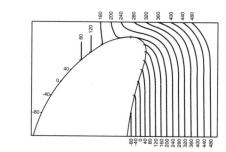

Ⓕ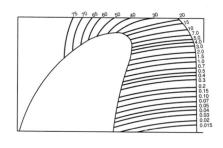

Name _Thomas O'Connell_   Date _____

Completion

Temp **1.** The pressure drop created by an expansion device is accompanied by a(n) _____ drop that enables the system to produce a cooling effect.

~~solid state~~ ~~sensor~~ **2.** A(n) _thermister_ is an electronic sensor used with a thermoelectric expansion valve.

heat **3.** A thermostatic expansion valve actually controls _____ in the refrigerant that leaves the evaporator coil of an air conditioning system.

distributor **4.** A(n) _____ splits refrigerant flow into several separate return bends on the evaporator coil to evenly distribute the refrigerant into the coils.

~~thermoelectr~~ **5.** A(n) _____ expansion valve is an expansion device that controls the flow of refrigerant in response to temperature sensed by a solid-state sensor.

flat plate **6.** Three types of coils used in air-cooled evaporators are bare-tube, finned-tube, and _____ coils.

suction **7.** The _____ line is the refrigerant line that runs from the evaporator coil to the compressor.

hand valve **8.** A(n) _____ expansion valve is a valve that opens or closes by the pressure in the refrigerant line ahead of the valve.

Cooler **9.** A(n) _____ is an evaporator that chills water. _aka "Chiller"_

cooling **10.** _____ capacity is the total heat transfer capacity of an evaporator coil expressed in Btu per hour.

True-False

T (F) **1.** A capillary tube expansion device can handle significant changes in the cooling load of a refrigeration system.

(T) F **2.** A float valve controls liquid level.

T (F) **3.** An air conditioner with a hand valve as an expansion device requires an operator to be present during use.

(T) F **4.** An automatic expansion valve is appropriate for an air conditioning system that has a continuously changing load.

(T) F **5.** The main advantage of using a thermostatic expansion valve is that it adjusts the refrigerant flow to match the cooling load on the system.

(T) F **6.** In a tube-and-shell cooler, the refrigerant in the system flows through the tubes and the water flows through the shell.

T̶ F **7.** In both tube-and-shell and coil-and-shell evaporators, the thickness of a copper tube or shell is all that separates the refrigerant from water.

T̶ F **8.** Expansion devices control the flow of refrigerant by controlling the volume, pressure, or ~~enthalpy~~ of the refrigerant in the low-pressure side of an air conditioning or refrigeration system.

T̶ F **9.** The thermostatic expansion valve and the automatic expansion valve control the ~~temperature~~ *pressure* of the refrigerant in a system.

(T) F **10.** The expansion device and compressor suction pressure maintain the low pressure in the low-pressure side of a mechanical compression refrigeration system.

Short Answer

1. What is the main reason for using a capillary tube as an expansion device?

refrigerant can flow through while system is off (cheap also)

2. Describe a tube-and-shell cooler.

cool that contains tubes, refrigerant flows through the tubes.

3. Describe the function of a suction accumulator.

prevents liquid from entering compressor

4. Describe the heat transfer process in an air-cooled evaporator.

heat travels from ambient air to the cooler refrigerant filled coils

5. What is the main difference between a chiller and an air conditioner?

chillers chill water opposed to air.

Compression System— Low-Pressure Side

Name _____   Date _____

Activity — Volume Control

Equip a refrigeration unit or trainer with a capillary tube expansion device. The unit should have service ports on the high- and low-pressure sides and a refrigerant flow gauge. Connect a manifold set to the unit and turn the unit ON. Record the high- and low-pressure side pressures and the refrigerant flow rate.

_____ **1.** The high-pressure side pressure is _____ psig.

_____ **2.** The low-pressure side pressure is _____ psig.

_____ **3.** The refrigerant flow rate is _____ fpm.

Replace the capillary tube with a tube one half the length of the original. Record the high- and low-pressure side pressures and the refrigerant flow rate.

_____ **4.** The high-pressure side pressure is _____ psig.

_____ **5.** The low-pressure side pressure is _____ psig.

_____ **6.** The refrigerant flow rate is _____ fpm.

Replace the capillary tube with a longer tube than the original. Record the high- and low-pressure side pressures and the refrigerant flow rate.

_____ **7.** The high-pressure side pressure is _____ psig.

_____ **8.** The low-pressure side pressure is _____ psig.

_____ **9.** The refrigerant flow rate is _____ fpm.

10. If the expansion device is a hand valve, how does the refrigerant flow rate change?

Activity — Pressure Control

Equip a refrigeration unit or trainer with an automatic expansion valve. The unit should have service ports on the high- and low-pressure sides and a refrigerant flow gauge. Connect a manifold set to the unit and turn the unit ON. Record the high- and low-pressure side pressures and the refrigerant flow rate.

_____ **1.** The high-pressure side pressure is _____ psig.

_____ **2.** The low-pressure side pressure is _____ psig.

_____ **3.** The refrigerant flow rate is _____ fpm.

Tighten the adjustment screw on the automatic expansion valve three turns. Record the high- and low-pressure side pressures and the refrigerant flow rate.

_____ **4.** The high-pressure side pressure is _____ psig.

_____ **5.** The low-pressure side pressure is _____ psig.

_____ **6.** The refrigerant flow rate is _____ fpm.

Loosen the adjustment screw on the automatic expansion valve six turns. Record the high- and low-pressure side pressures and the refrigerant flow rate.

_____ **7.** The high-pressure side pressure is _____ psig.

_____ **8.** The low-pressure side pressure is _____ psig.

_____ **9.** The refrigerant flow rate is _____ fpm.

10. Did the flow rate change when the automatic expansion valve was adjusted? Explain.

Activity — Temperature Control

Equip a refrigeration unit or trainer with a thermostatic expansion valve. The unit should have service ports on the high- and low-pressure sides and a refrigerant flow gauge. Connect a manifold set to the unit and attach a thermometer to the suction line where it leaves the evaporator coil. Turn the unit ON. Record the high- and low-pressure side pressures, refrigerant flow rate, temperature of the refrigerant leaving the evaporator coil, and superheat.

_____ **1.** The high-pressure side pressure is _____ psig.

_____ **2.** The low-pressure side pressure is _____ psig.

_____ **3.** The refrigerant flow rate is _____ fpm.

_____ **4.** The temperature of the refrigerant leaving the evaporator coil is _____ °F.

_____ **5.** The superheat is _____ °F.

Tighten the adjustment screw on the thermostatic expansion valve three turns. Record the high- and low-pressure side pressures and refrigerant flow rate.

_____ **6.** The high-pressure side pressure is _____ psig.

_____ **7.** The low-pressure side pressure is _____ psig.

_____ **8.** The refrigerant flow rate is _____ fpm.

Loosen the adjustment screw on the thermostatic expansion valve six turns. Record the high- and low-pressure side pressures and the refrigerant flow rate.

_____   **9.** The high-pressure side pressure is _____ psig.

_____   **10.** The low-pressure side pressure is _____ psig.

_____   **11.** The refrigerant flow rate is _____ fpm.

_____   **12.** Did the superheat in the system increase or decrease when the thermostatic expansion valve adjustment screw was tightened?

Activity — Air Conditioner Cooling Effect

Equip a refrigeration unit or trainer with any expansion device. The unit should have service ports on the high- and low-pressure sides and a refrigerant flow gauge. Install a manifold set on the service ports. Attach thermometers on the liquid line just ahead of the expansion device and on the suction line where it leaves the evaporator coil. Also install thermometers in the air inlet and outlet of the evaporator coil. Turn the unit ON. Record the high- and low-pressure side pressures, refrigerant flow rate, refrigerant temperature entering the expansion device, temperature of the refrigerant leaving the evaporator coil, and temperature of the air entering and leaving the evaporator coil.

_____   **1.** The high-pressure side pressure is _____ psig.

_____   **2.** The low-pressure side pressure is _____ psig.

_____   **3.** The refrigerant flow rate is _____ fpm.

_____   **4.** The refrigerant temperature entering the expansion device is _____°F.

_____   **5.** The refrigerant temperature leaving the evaporator coil is _____°F.

_____   **6.** The air temperature entering the evaporator is _____°F.

_____   **7.** The air temperature leaving the evaporator is _____°F.

Use a pressure-enthalpy diagram or refrigerant property table for the refrigerant to calculate the refrigeration effect of the unit. Refer to the appropriate pressure-enthalpy diagram or refrigerant property table on pages 173–187.

_____   **8.** The refrigeration effect of the unit is _____ Btu/lb.

Calculate the cooling capacity of the unit using the refrigeration effect and the mass flow rate of the refrigerant.

_____   **9.** The cooling capacity calculated from the refrigeration effect and the mass flow rate of the refrigerant is _____ Btu/hr.

Calculate the cooling capacity of the unit using the temperature drop of the air across the evaporator coil and an estimate of the volumetric flow rate of air across the evaporator coil.

_____   **10.** The cooling capacity calculated from the temperature drop of the air across the evaporator coil and the volumetric flow rate of air across the evaporator coil is _____ Btu/hr.

_____ **11.** Should any difference exist?

12. If a difference exists, what variable caused it?

Activity — Chiller Cooling Effect

Equip a chiller unit or trainer with any expansion device. The unit should have service ports on the high- and low-pressure sides and a refrigerant flow gauge. Install a manifold set on the service ports. Attach thermometers on the liquid line just ahead of the expansion device and on the suction line where it leaves the cooler. Also install thermometers on the cooler water connections. Turn the unit ON. Record the high- and low-pressure side pressures, refrigerant flow rate, refrigerant temperature entering the expansion device, temperature of the refrigerant leaving the cooler, and temperature of the water entering and leaving the cooler.

_____ **1.** The high-pressure side pressure is _____ psig.

_____ **2.** The low-pressure side pressure is _____ psig.

_____ **3.** The refrigerant flow rate is _____ fpm.

_____ **4.** The refrigerant temperature entering the expansion device is _____°F.

_____ **5.** The refrigerant temperature leaving the cooler is _____°F.

_____ **6.** The water temperature entering the cooler is _____°F.

_____ **7.** The water temperature leaving the cooler is _____°F.

Use a pressure-enthalpy diagram or refrigerant property table for the refrigerant to calculate the refrigeration effect of the unit. Refer to the appropriate pressure-enthalpy diagram or refrigerant property table on pages 173–187.

_____ **8.** The refrigeration effect of the unit is _____ Btu/lb.

Calculate the cooling capacity of the unit using the refrigeration effect and the mass flow rate of the refrigerant.

_____ **9.** The cooling capacity calculated from the refrigeration effect and the mass flow rate of the refrigerant is _____ Btu/hr.

Calculate the cooling capacity of the unit using the temperature drop of the water through the cooler and an estimate of the volumetric flow rate of water through the cooler.

_____ **10.** The cooling capacity calculated from the temperature drop of the water through the cooler and the volumetric flow rate of water through the cooler is _____ Btu/hr.

_____ **11.** Should any difference exist?

12. If a difference exists, what variable caused it?

Name _____ Date _____

Calculations

Refer to Refrigerant Property Tables on pages 173–187.

_____ 1. A refrigeration system containing R-407c has a low-pressure side pressure of 85.6 psig. The mass flow rate of the refrigerant is 19.6 lb/min. The cooling capacity of the evaporator is _____ Btu/hr.

_____ 2. The cooling capacity of an air-cooled evaporator is 140,000 Btu/hr. The actual refrigeration effect of the system is _____ Btu/hr.

_____ 3. An air-cooled evaporator produces a refrigeration effect of 92.6 Btu/lb. The mass flow rate of refrigerant is 27.4 lb/min. The cooling capacity of the evaporator is _____ Btu/hr.

_____ 4. A refrigeration system containing R-22 has a low-pressure side pressure of 111.26 psig. The mass flow rate of the refrigerant is 22.5 lb/min. The cooling capacity of the evaporator is _____ Btu/hr.

_____ 5. An air-cooled refrigeration system containing R-12 has a low-pressure side pressure of 70.2 psig. The mass flow rate of the refrigerant is 17 lb/min. The actual refrigeration effect of the system is _____ Btu/hr.

_____ 6. The cooling capacity of a cooler is 480,000 Btu/hr. The actual refrigeration effect of the cooler is _____ Btu/hr.

_____ 7. On a refrigerant property table for R-134a, the enthalpy of the refrigerant as a vapor is 116.3 Btu/lb at 124.3 psig and the enthalpy of the refrigerant as a liquid is 45.1 Btu/lb at 124.3 psig. The refrigeration effect is _____ Btu/lb.

_____ 8. A cooler produces a refrigeration effect of 65.3 Btu/lb. The mass flow rate of the refrigerant is 14 lb/min. The cooling capacity of the cooler is _____ Btu/hr.

_____ 9. On a refrigerant property table for R-502, the enthalpy of the refrigerant as a vapor is 77.3 Btu/lb at 27.5 psig and the enthalpy of the refrigerant as a liquid is 8.6 Btu/lb at 27.5 psig. The refrigeration effect is _____ Btu/lb.

_____ 10. A chiller containing R-12 has a low-pressure side pressure of 46.74 psig. The mass flow rate of the refrigerant is 21 lb/min. The actual refrigeration effect of the system is _____ Btu/hr.

_____ 11. A cooler produces a refrigeration effect of 87.5 Btu/lb. The mass flow rate of the refrigerant is 26.3 lb/min. The cooling capacity of the cooler is _____ Btu/hr.

_____ 12. On a refrigerant property table for R-22, the enthalpy of the refrigerant as a vapor is 106.65 Btu/lb at 48.8 psig and the enthalpy of the refrigerant as a liquid is 17.3 Btu/lb at 48.8 psig. The refrigeration effect is _____ Btu/lb.

Short Answer

1. Define service valve.

2. Define sight glass.

3. Define moisture indicator.

4. Define filter-dryer.

Identification

_____ **1.** Evaporator
_____ **2.** Accessories
_____ **3.** Blower
_____ **4.** Cabinet

Name _Thomas O'Connell_   Date _____

Short Answer

1. What is the main function of the compressor in an air conditioning system?

   Pump refrigerant/increase pressure.

2. What is the main function of the condenser in an air conditioning system?

   reject heat/condense liquid refrigerant

3. What is a positive-displacement compressor?

   has a fixed amount of refrigerant moving during each cycle (what goes in comes out)

4. What is the main function of the liquid receiver in an air conditioning system?

   prevent compressor damage (* stores liquid refrigerant)

5. Why is an evaporative condenser more efficient at transferring heat than air- or water-cooled condensers?

   a large amount of heat is removed by evaporation. (uses latent heat)

True-False

T   F   1. A centrifugal compressor works on the same principle as a centrifugal air blower.

T   F   2. As refrigerant leaves the impeller wheel of a centrifugal compressor, the speed of the refrigerant is converted to pressure because the refrigerant is forced into a smaller opening.

T   F   3. A centrifugal compressor is used for applications that require high pressure and low volume of refrigerant flow.

T   F   4. As a refrigerant flows through a condenser coil, heat moves from the air moving across the coil to the refrigerant flowing through the coil.

T   F   5. One pipe sizing chart is used for sizing refrigerant lines for refrigeration systems.

T̶ F 6. During the suction stroke of a reciprocating compressor, refrigerant vapor flows into the cylinder because the pressure in the cylinder is lower than the pressure of the refrigerant in the suction line.

(T) F 7. During the compression stroke of a reciprocating compressor, the pressure in the cylinder increases and closes the suction valve.

(T) F 8. In a shell-and-tube condenser, water flows through the shell and refrigerant flows through the tubes.

T (F) 9. A semi-hermetic compressor is serviced on a job site.

(T) F 10. To achieve higher pressures with centrifugal compressors, multistage units are used.

Multiple Choice

_d_ 1. Compressors classified by mechanical action include all of the following except _____ compressors.

A. reciprocating
B. centrifugal
C. rotary
D. open

_d_ 2. _____ water-cooled condensers are used in water-cooled refrigeration systems.

A. Double pipe
B. Shell-and-coil
C. Shell-and-tube
D. all of the above

_a_ 3. All of the following except _____ condensers are basic condensers used on refrigeration systems.

A. atmospheric
B. air-cooled
C. evaporative
D. water-cooled

_a_ 4. The most commonly used accessory on the high-pressure side of a refrigeration system is the _____.

A. filter-dryer
B. distributor
C. liquid receiver
D. thermostatic expansion valve

_a_ 5. All of the following factors except the _____ determine the volumetric capacity of a compressor.

A. capacity of each cylinder
B. speed of compressor
C. type of refrigerant
D. number of cylinders

Completion

pistons 1. In a reciprocating compressor, _____ move back and forth in closed cylinders.

rotating 2. Two kinds of rotary compressors used in air conditioning systems are stationary vane and _____ vane compressors.

plate 3. Leaf valves and _____ valves are used in reciprocating compressors.

rotary _screw_ 4. A(n) _____ compressor compresses refrigerant with two screw-like helical gears that interlock as they turn.

hermetic 5. A(n) _____ compressor is a compressor that has all of the components and the motor sealed in a metal housing.

13 Compression System– High-Pressure Side

Activities

Name _____   Date _____

Activity — Open Compressor

Obtain an open compressor that can be scrapped. Open the service valves on the compressor or carefully loosen the refrigerant line connections to make sure no refrigerant or pressure is in the compressor. If the compressor has a drain plug, drain the oil. If the compressor does not have a drain plug, drain the oil when the unit is open. Handle the oil carefully, as it may be acidic. Dispose of the oil properly. Dismantle the compressor. Keep track of all fasteners.

1. Trace the refrigerant path as it enters and leaves the compressor.
2. Locate the suction and discharge valves.
3. Explain the mechanical operation of the compressor.

4. Locate and identify the seal used on the shaft extension.

Reassemble the compressor.

Activity — Semi-hermetic Compressor

Obtain a semi-hermetic compressor that can be scrapped. Open the service valves on the compressor or carefully loosen the refrigerant line connections to make sure no refrigerant or pressure is in the compressor. If the compressor has a drain plug, drain the oil. If the compressor does not have a drain plug, drain the oil when the unit is open. Handle the oil carefully, as it may be acidic. Dispose of the oil properly. Dismantle the compressor. Keep track of all

1. Trace the refrigerant path as it enters and leaves the compressor.
2. Locate the suction and discharge valves.

3. Explain the mechanical operation of the compressor.

4. Describe the lubrication system.

5. Describe the motor drive.

Reassemble the compressor.

Activity — Reciprocating Compressor

Obtain a dismantled reciprocating compressor. Examine the

        1. Explain how compression is achieved.

        2. Identify the crank shaft, connecting rods, wrist pins, pistons, and oil pump.

_____ **3.** Determine the number of cylinders.

Measure the bore of the cylinders and stroke of the pistons.

_____ **4.** The bore of the cylinders is _____″.

_____ **5.** The stroke of the pistons is _____″.

        6. How is mechanical motion transferred from the crankshaft to the pistons?

        7. How are the moving parts lubricated?

        8. What is the displacement of one piston?

        9. What is the displacement of the compressor?

Activity — Rotary Compressor

Obtain a dismantled rotary compressor. Examine the compressor.

1. Explain how compression is achieved.

2. Identify the cylinder, rotating piston, seals, discharge valve, and oil pump.
3. How are the compressor shaft and motor connected?

4. How is the seal between the piston and cylinder walls maintained?

5. How are the moving parts lubricated?

Activity — Screw Compressor

Obtain a dismantled screw compressor. Examine the compressor.

1. Explain how compression is achieved.

2. Identify the crank shaft, interlocking screws, intake valve, discharge valve, capacity control valve, and oil pump.
3. Are the interlocking screws identical? Explain.

4. How are the screws turned?

5. How is refrigerant flow controlled?

Activity — Heat of Rejection

Equip an air-cooled refrigeration unit or trainer with an expansion device. The unit should have service ports on the high- and low-pressure sides and a refrigerant flow gauge. Install a manifold set on the service ports. Attach thermometers on the refrigerant lines entering and leaving the condenser coil. Also install thermometers in the air inlet and outlet of the condenser coil. Turn the unit ON. Record the high- and low-pressure side pressures, refrigerant flow rate, temperature of the refrigerant entering and leaving the condenser coil, and temperature of the air entering and leaving the condenser coil.

_____ **1.** The high-pressure side pressure is _____ psig.

_____ **2.** The low-pressure side pressure is _____ psig.

_____ **3.** The refrigerant flow rate is _____ fpm.

_____ **4.** The temperature of the refrigerant entering the condenser coil is _____°F.

_____ **5.** The temperature of the refrigerant leaving the condenser coil is _____°F.

_____ **6.** The temperature of the air entering the condenser coil is _____°F.

_____ **7.** The temperature of the air leaving the condenser coil is _____°F.

Use a pressure-enthalpy diagram or refrigerant property table for the refrigerant to calculate the heat of rejection. Refer to the appropriate pressure-enthalpy diagram or refrigerant property table on pages 173–187.

_____ **8.** The heat of rejection is _____ Btu/lb.

Calculate the heat rejection rate of the unit using the heat of rejection and the mass flow rate of the refrigerant.

_____ **9.** The heat rejection rate calculated from the heat of rejection and the mass flow rate of the refrigerant is _____ Btu/hr.

Calculate the heat rejection rate of the unit using the temperature increase of the air across the condenser coil and an estimate of the volumetric flow rate of air across the condenser coil.

_____ **10.** The heat rejection rate calculated from temperature increase of the air across the condenser coil and the volumetric flow rate of air across the condenser coil is _____ Btu/hr.

_____ **11.** Should any difference exist?

12. If a difference exists, what variable caused it?

Compression System– High-Pressure Side

Trade Test

Name _____   Date _____

Calculations

_____ **1.** The pistons in a compressor have diameters of 1.75″ and stroke lengths of 1.45″. The volumetric capacity of one cylinder is _____ cu in.

_____ **2.** A refrigeration system that contains R-22 has a low-pressure side pressure of 84 psig and a high-pressure side pressure of 168 psig. The mass flow rate of the refrigerant that provides a 100,000 Btu/hr cooling capacity is _____ lb/min.

_____ **3.** A compressor has six cylinders. The pistons in the cylinders have diameters of 2″ and stroke lengths of 1.75″. What is the volumetric capacity of the compressor if it turns at 1250 rpm?

_____ **4.** A refrigeration system has a mass flow rate of 22.56 lb/min. The specific volume of the refrigerant is .5545 cu ft/lb. The volumetric flow rate of the refrigerant is _____ cfm.

_____ **5.** A refrigeration system that contains R-12 has a low-pressure side pressure of 117 psig and a high-pressure side pressure of 233 psig. The mass flow rate of the refrigerant that provides a 60,000 Btu/hr cooling capacity is _____ lb/min.

_____ **6.** The pistons in a compressor have diameters of 1.5″ and stroke lengths of 1.25″. The volumetric capacity of one cylinder is _____ cu in.

_____ **7.** A refrigeration system that contains R-502 has a high-pressure side pressure of 201.4 psig. The mass flow rate of the refrigerant is 22.3 lb/min. The heat rejection rate of the condenser is _____ Btu/hr.

_____ **8.** A compressor has two cylinders. The pistons in the cylinders have diameters of 1.25″ and stroke lengths of 1.5″. If the compressor turns at 1750 rpm, the volumetric capacity of the compressor is _____ cfm.

_____ **9.** A condenser coil has a heat rejection rate of 57,500 Btu/hr. The condenser blower moves 2600 cfm of air across the coil. The temperature increase of the air leaving the condenser coil is _____°F.

_____ **10.** A refrigeration system that contains R-12 has a water-cooled condenser at a high-pressure side pressure of 233.73 psig. The mass flow rate of the refrigerant is 26.2 lb/min. The heat rejection rate of the condenser is _____ Btu/hr.

_____ **11.** A refrigeration system has a mass flow rate of 19.72 lb/min. The specific volume of the refrigerant is .27 cu ft/lb. The volumetric flow rate of the refrigerant is _____ cfm.

_____ **12.** A water-cooled condenser has a heat rejection rate of 67,824 Btu/hr. The volumetric flow rate of water through the condenser is 17 gpm. The temperature increase of the water leaving the condenser is _____°F.

_____ **13.** A refrigeration system that contains R-22 has a high-pressure side pressure of 296.96 psig. The mass flow rate of the refrigerant is 18.9 lb/min. The heat rejection rate of the condenser is _____ Btu/hr.

_____ **14.** A condenser coil has a heat rejection rate of 108,000 Btu/hr. The condenser blower moves 4000 cfm of air across the coil. The temperature increase of the air leaving the condenser coil is _____ °F.

_____ **15.** A refrigeration system has a water-cooled condenser with a water flow rate of 17 gpm. The temperature of the water entering the condenser is 56°F and the temperature of the water leaving the condenser is 67°F. The heat rejection rate of the water is _____ Btu/hr.

Identification — Air-Cooled Condenser

_____ **1.** Fan

_____ **2.** Condenser coil

_____ **3.** Compressor

_____ **4.** Cabinet

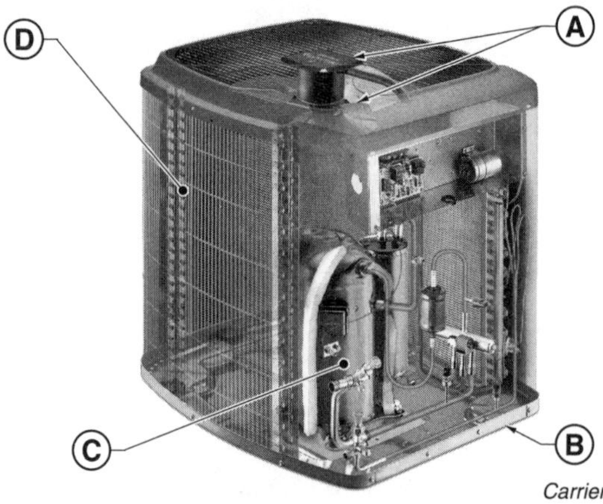

Carrier Corporation

Identification — Water-Cooled Condenser

_____ **1.** Frame

_____ **2.** Condenser coil

_____ **3.** Compressor

_____ **4.** Refrigerant line

_____ **5.** Accessories

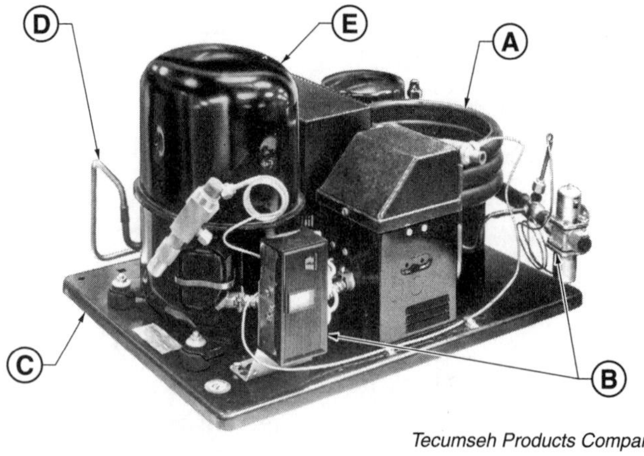

Tecumseh Products Company

Name _Thomas O'Connell_   Date _____

True-False

T **(F)**   **1.** Package air conditioning units are always located on the ground outside a building.

T **(F)**   **2.** The cooling anticipator in a cooling thermostat produces heat when the switch is in the open position.

(T) F   **3.** A four-pipe hydronic distribution system allows hot or cold water to flow to any terminal device.

T F   **4.** A circulating pump on a hydronic air conditioning system removes condensate from the system.

T F   **5.** All terminal devices used in hydronic air conditioning systems must have forced-air circulation.

T F   **6.** A combination unit is an air conditioner that contains the components for cooling and heating in one sheet metal cabinet.

T F   **7.** Duct coils may be supplied with air from a remotely located blower.

T F   **8.** Because air has a low density and low specific heat, large ductwork must be used.

(T) F   **9.** A cooling tower is an evaporative heat exchanger that removes heat from air.

T **(F)**   **10.** A split system is an air conditioning system that has separate cabinets for the evaporator and condenser.

Completion

_____   **1.** Three types of cooling towers used with water-cooled air conditioning systems are forced draft, induced draft, and _____ cooling towers.

_____   **2.** The two most common evaporating mediums used in air conditioning systems are water and _____.

_____   **3.** A unit _____ cools and ventilates a room simultaneously.

_____   **4.** A(n) _____ is the component in a hydronic air conditioning system that cools water, which cools air.

_____   **5.** A(n) _____ unit is a self-contained air conditioner that has all of the components contained in one sheet metal cabinet.

_____   **6.** _____ condensers use air and water as condensing mediums.

_____ 7. Air conditioners use the mechanical compression or the _____ refrigeration process.

_____ 8. A(n) _____ pressure switch ensures that there is lubricating oil in the system when the system is operating.

_____ 9. The three categories of controls found on air conditioning systems are power, operating, and _____ controls.

_____ 10. In a forced air conditioning system, heat is transferred between air and refrigerant in the evaporator and _____.

_____ 11. _____ air conditioning systems are used where the air conditioning equipment is located centrally and the building spaces are located remotely.

_____ 12. _____ systems distribute water from the chiller to the terminal devices.

Short Answer

1. Define condensing medium.

2. List and describe two types of supply duct systems used with a combination unit.

3. Define evaporating medium.

4. What two factors make water a better heating medium than air?

5. What is the function of the transformer in an air conditioning system?

6. What is the function of pressure switches on air conditioning systems?

7. What is the difference between a contactor and a magnetic starter?

8. Define nominal size.

Name _____   Date _____

Activity — Inspection

Inspect the air conditioning system on a building site or manufacturing complex after getting permission from an authority at the site. Complete the inspection report.

Date _____   Site _____

Person in charge _____   Permission granted by _____

Equipment manufacturer _____   Refrigerant _____

Type of system

Electrical characteristics

Activity — Identification

Identify the components of a forced-air air conditioning system.

_____   **1.** Ductwork

_____   **2.** Controls

_____   **3.** Register

_____   **4.** Blower

_____   **5.** Grill

_____   **6.** Air conditioner

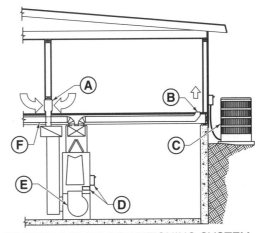

FORCED-AIR AIR CONDITIONING SYSTEM

Identify the components of a hydronic air conditioning system.

_____ **7.** Controls

_____ **8.** Chiller

_____ **9.** Terminal devices

_____ **10.** Piping

_____ **11.** Circulating pump

HYDRONIC AIR CONDITIONING SYSTEM

Activity — Piping Systems

Draw the piping for a three-pipe hydronic distribution system by connecting the parts with lines. Mark the direction of water flow on each section of pipe with arrows.

HOT SUPPLY PIPING	————————
COLD SUPPLY PIPING	————————
RETURN PIPING	-------------------------

BOILER

CHILLER

Draw the piping for a four-pipe hydronic distribution system by connecting the parts with lines. Mark the direction of water flow on each section of pipe with arrows.

HOT SUPPLY PIPING	————————
COLD SUPPLY PIPING	————————
HOT RETURN PIPING	-------------------------
COLD RETURN PIPING	-------------------------

BOILER

CHILLER

Air Conditioning Systems

Trade Test

Name _____ Date _____

Calculations

_____ **1.** A 650 ton chiller produces _____ Btu/hr of cooling.

_____ **2.** A 6 ton air conditioner produces _____ Btu/hr of cooling.

_____ **3.** A refrigeration system that produces 118,000 Btu/hr of cooling has a nominal size of _____ tons.

_____ **4.** How much heat is required to raise 10 lb of air 1°F?

_____ **5.** A .5 ton air conditioner produces _____ Btu/hr of cooling.

_____ **6.** A ton of cooling equals _____ Btu/hr.

_____ **7.** A 7.5 ton air conditioner produces _____ Btu/hr of cooling.

_____ **8.** A refrigeration system that produces 23,000 Btu/hr of cooling has a nominal size of _____ tons.

_____ **9.** How much heat is required to raise 15 lb of water 1°F?

_____ **10.** A 33 ton air conditioner produces _____ Btu/hr of cooling.

Short Answer

1. Explain the function of a disconnect in an air conditioning system.

2. Explain ST and DT as related to electromechanical relays.

3. Explain the function of fuses in air conditioning electrical circuits.

4. Explain how electromechanical relays are identified.

5. Explain NO and NC as related to electromechanical relays.

6. What is the function of overload relays in a magnetic starter?

7. Explain how a cooling thermostat differs from a heating thermostat.

8. Explain DP and SP as related to electromechanical relays.

Identification

_____ **1.** Evaporator coil

_____ **2.** Blower

_____ **3.** Compressor

_____ **4.** Condenser coil

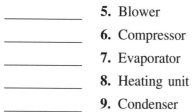

_____ **5.** Blower

_____ **6.** Compressor

_____ **7.** Evaporator

_____ **8.** Heating unit

_____ **9.** Condenser

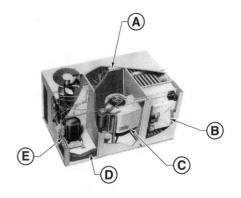

15 Heat Pumps

Name _Thomas O'Connell_   Date _____

Short Answer

1. Define heat sink.

   Heat sink is a substance with cold surface capable of absorbing heat.

2. Define heat pump.

   a mechanical compression refrigeration system that moves heat from one area to another.

3. Explain the defrost cycle of a heat pump.

   reverses refrigerant flow.

4. Describe a heat pump thermostat.

   incorporates system switch, heat thermostat, and cooling thermostat.

5. Describe a solenoid-operated pilot valve.

6. Describe a refrigerant control valve.

7. Why are defrost controls required on a heat pump and not on an air conditioning system?

8. The heat pump coil is located on the outlet side of an existing heating system. Why must the heat pump be shut down when the auxiliary heat turns ON?

True-False

T   F   **1.** In a water-to-air heat pump, heat is transferred from water to a refrigerant and then from the refrigerant to the air.

T   F   **2.** In the defrost cycle, a heat pump operates in the cooling mode when the outdoor blower is OFF.

T   F   **3.** As a heat pump switches from the cooling mode to the heating mode, the pressure in the two sides of the system remains the same.

T   F   **4.** The balance point temperature of a heat pump is the point at which the pressure of the refrigerant equals the outdoor temperature.

T   F   **5.** A solenoid moves the piston of a pilot valve.

T   F   **6.** A capillary tube can be used as an expansion device on a heat pump because it produces a pressure decrease in the refrigerant regardless of the direction of flow.

T   F   **7.** The size of the port in a refrigerant control valve is larger when the refrigerant flows one direction than it is when the refrigerant flows the other direction.

T   F   **8.** A thermostatic expansion valve allows refrigerant flow in two directions.

T   F   **9.** Auxiliary heat is used to prevent the compressor in a heat pump from running continuously.

T   F   **10.** A tapping valve is a valve that pierces a refrigerant line.

Completion

_____   **1.** The direction of the refrigerant flow in a heat pump is controlled by a(n) _____ valve.

_____   **2.** The two mediums most commonly used for heat sinks in heat pumps are air and _____.

_____   **3.** When a heat pump is in the cooling mode, the refrigerant leaves the reversing valve and flows to the _____ coil.

_____   **4.** A heat pump normally is sized to provide the heat required to heat a building to an outdoor air temperature of approximately _____°F.

_____   **5.** When thermostatic expansion valves are used at each coil in a heat pump, bypass piping with a(n) _____ valve in it must be installed.

_____   **6.** A reversing valve is operated by _____ pressure.

_____   **7.** The three types of expansion devices used on heat pumps are refrigerant control valves, _____ tubes, and thermostatic expansion valves.

_____   **8.** Most heat pumps require multibulb, _____ thermostats.

_____   **9.** Hot _____ provides the heat for defrosting an outdoor coil during a defrost cycle.

_____   **10.** Most defrost control systems are based on _____ cycles.

Heat Pumps

Activities

Name _____   Date _____

Activity — Heat Pump Performance Chart

Obtain a heat pump performance chart. Plot the heating output for the heat pump at different outdoor temperatures. Connect the points to establish the heat output of the pump over the temperature range.

_____ **1.** What is the heat output of the heat pump at an outdoor temperature of 45°F?

_____ **2.** If a building has a heat loss of 67,000 Btu/hr at 30°F, does the heat pump provide all the heat necessary for heating the building at that temperature? If not, how much additional heat is needed?

3. How does a heat pump output graph determine the balance point temperature for an application?

PUMP OUTPUT (BTU/HR)

OUTDOOR AIR TEMPERATURE (°F)

Activity — Reversing Valve

Equip a refrigeration unit or trainer with a capillary tube expansion device. Disconnect the refrigerant lines from the suction and discharge connections on the compressor and install a reversing valve in the system. Use an air conditioning thermostat to operate the unit. Install a low-voltage switch to energize the reversing valve solenoid.

_____ **1.** Does energizing the reversing valve solenoid cause the refrigerant to flow through the compressor in a different direction?

_____ **2.** Does the capillary tube work as an expansion device when the refrigerant flows in either direction?

Activity — Refrigerant Flow

Use arrows to identify the direction of refrigerant flow in each system.

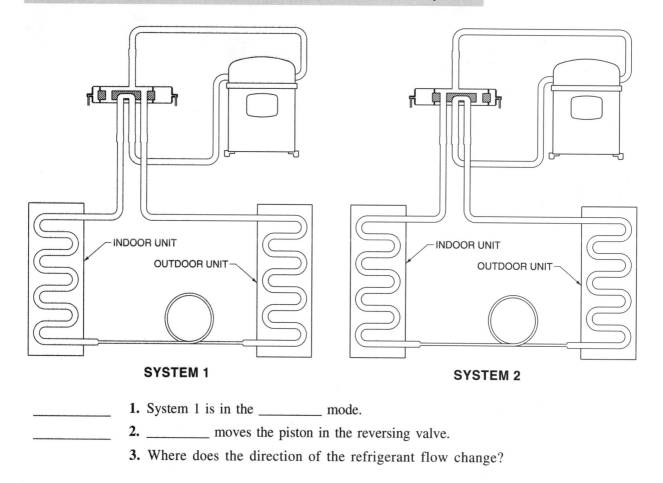

SYSTEM 1　　　　　　　　　**SYSTEM 2**

_____ **1.** System 1 is in the _____ mode.

_____ **2.** _____ moves the piston in the reversing valve.

3. Where does the direction of the refrigerant flow change?

Activity — Air-to-Air Heat Pump

Equip a refrigeration unit or trainer with a capillary tube expansion device. Disconnect the refrigerant lines from the suction and discharge connections on the compressor and install a reversing valve in the system. Use an air conditioning thermostat to operate the unit. Install a low-voltage switch to energize the reversing valve solenoid. Place thermometers in the air inlet and outlet of the two coils. Turn the unit ON and operate it in the cooling mode. Measure and record the temperature of the air entering and leaving the indoor and outdoor coils.

_____ **1.** The temperature of the air entering the indoor coil is _____ °F.

_____ **2.** The temperature of the air leaving the indoor coil is _____ °F.

_____ **3.** The temperature of the air entering the outdoor coil is _____ °F.

_____ **4.** The temperature of the air leaving the outdoor coil is _____ °F.

Calculate the temperature decrease through the indoor coil and the temperature increase through the outdoor coil.

_____   **5.** The temperature decrease through the indoor coil is _____°F.

_____   **6.** The temperature increase through the outdoor coil is _____°F.

While the unit is operating, switch the reversing valve by energizing the reversing valve solenoid. The unit should switch to the heating mode. After several minutes, measure and record the temperature of the air entering and leaving the indoor and outdoor coils.

_____   **7.** The temperature of the air entering the indoor coil is _____°F.

_____   **8.** The temperature of the air leaving the indoor coil is _____°F.

_____   **9.** The temperature of the air entering the outdoor coil is _____°F.

_____   **10.** The temperature of the air leaving the outdoor coil is _____°F.

Calculate the temperature increase through the indoor coil and the temperature decrease through the outdoor coil.

_____   **11.** The temperature increase through the indoor coil is _____°F.

_____   **12.** The temperature decrease through the outdoor coil is _____°F.

Switch the reversing valve back to the cooling mode before shutting the unit OFF.

13. How does a heat pump operating in both cooling and heating modes demonstrate the fact that an air conditioner transfers heat?

Activity — Air-to-Water Heat Pump

Replace the coil used as the indoor coil in the cooling mode with a copper coil. Submerge the coil in a container of water. The water should be at room temperature. Place a thermometer in the water. Measure and record the temperature of the water.

_____   **1.** The initial water temperature is _____°F.

Operate the unit in the cooling mode for 1 hour or until the water temperature drops to approximately 35°F. Measure and record the temperature of the water.

_____   **2.** The final water temperature is _____°F.

Calculate and record the difference between the initial and final water temperature.

_____   **3.** The water temperature difference is _____°F.

4. What happened to the temperature of the water?

5. What caused the temperature difference?

6. What happened to the heat that was extracted from the water?

Replace the water in the container with room temperature water. Measure and record the temperature of the water.

_____ **7.** The initial water temperature is _____°F.

Operate the unit in the heating mode for 1 hour. Measure and record the temperature of the

_____ **8.** The final water temperature is _____°F.

Calculate and record the difference between the initial and final water temperature.

_____ **9.** The water temperature difference is _____°F.

10. What happened to the temperature of the water?

11. What caused the temperature difference?

12. Where did the heat that caused the difference come from?

Name _____   Date _____

Short Answer

1. Why does the amount of heat available from outdoor air vary?

2. What activates a defrost control system?

3. How many different combinations of heat pumps are possible considering the most commonly used heat sink mediums and the cooling/heating mediums?

4. Why is it incorrect to refer to the two coils in a heat pump as the evaporator and condenser coils?

5. What deactivates a defrost control system?

6. Explain why refrigerant pressure can determine coil temperature.

7. Why can a heat pump be called a reverse cycle air conditioning system?

8. In the heating mode, what operating characteristic of a heat pump makes a defrost function necessary?

9. Name the three devices in a heat pump that are affected when a heat pump goes into defrost. Describe how each is affected.

10. Why is a bypass circuit and a check valve located at each expansion valve when thermostatic expansion valves are used as expansion devices on a heat pump?

11. Why is auxiliary heat necessary with a heat pump?

12. Why are filters required at each end of a capillary tube when used as an expansion device on a heat pump?

Name _____ Date _____

True-False

T	F	**1.**	Power controls are installed in an electrical circuit between the power source and the heating unit.
T	F	**2.**	Temperature limit switches are used to start the burner on a heating unit.
T	F	**3.**	A flame rod is an electronic combustion safety control that consists of a light-sensitive device that detects flame.
T	F	**4.**	A water temperature control system is the group of components that control the temperature of water in a hydronic heating system.
T	F	**5.**	Central supervisory control systems do not control a load directly but enable or disable local controllers.
T	F	**6.**	A field interface device is an electronic device that sends commands to the CPU of a central-direct digital control system.
T	F	**7.**	A digital (binary) input device is a device that produces an ON or OFF signal.
T	F	**8.**	A flow switch is a device that surrounds a wire and detects the electromagnetic field caused by electricity passing through the wire.
T	F	**9.**	A pneumatic control system uses water pressure as the control signal.
T	F	**10.**	A remote bulb thermostat used to sense temperature contains a mechanism that opens and closes the contacts according to changes in pressure.

Completion

_____ **1.** Overcurrent protection devices such as fuses or circuit breakers are placed in an electrical circuit to protect the circuit from _____.

_____ **2.** A flame rod is an electronic combustion safety control that uses the _____ as an electrical conductor.

_____ **3.** A(n) _____ is an accordion-like device that converts pressure variation into mechanical movement.

_____ **4.** A(n) _____ package is the damper and controls that bring outdoor air indoors to cool building spaces.

_____ **5.** Commercial control system devices are often arranged in a closed loop configuration that provides _____ between the controlled device and controller.

_____ 6. Electronic control systems use _____ devices, such as diodes, LEDs, transistors, and thyristors, to control HVAC units.

_____ 7. A(n) _____ is a controller that coordinates communications from controller to controller on a network and provides a place for operator interface.

_____ 8. A(n) _____ input is a device that senses a variable such as temperature, pressure, or humidity, and causes a proportional electrical signal change at the controller.

_____ 9. A pneumatic _____ is a device, mounted to a damper or valve, that allows an air pressure signal to cause a mechanical motion.

_____ 10. A schematic diagram uses lines and _____ to represent the electrical circuits and components in an electrical control system.

Multiple Choice

_____ 1. _____ controls cycle equipment ON or OFF.
A. Safety
B. Power
C. Operating
D. all of the above

_____ 2. Most residential control circuit transformers are step-down transformers with a secondary voltage of _____.
A. 120 VAC
B. 120 VDC
C. 24 VAC
D. 24 VDC

_____ 3. Relays use a _____ to create magnetism to open or close one or more sets of contacts.
A. coil
B. thermostat
C. fuse
D. transformer

_____ 4. A(n) _____ is a device that contains a humidity-sensing element that changes characteristics with changes in humidity.
A. humidistat
B. pressurestat
C. thermostat
D. Aquastat®

_____ 5. A(n) _____ controller is designed to control only one type of HVAC system.
A. application-specific
B. universal input-output
C. network communication module
D. all-in-one

_____ 6. A building automation system _____ changes the state of a controlled device in response to a command from the controller.
A. input device
B. output device
C. transistor
D. thyristor

_____ 7. Three types of control systems are electrical, electronic, and _____.
A. automatic
B. manual
C. fluidic
D. pneumatic

_____   **8.** A flame surveillance combustion safety control senses _____.

    A. sound   C. temperature

    B. pressure   D. light

_____   **9.** A stack switch senses _____.

    A. carbon dioxide levels   C. flue gas temperature

    B. carbon monoxide levels   D. presence of smoke

_____   **10.** A thermistor is a sensing element that changes electrical resistance in response to a change in _____.

    A. temperature   C. humidity

    B. pressure   D. air flow rate

Short Answer

1. Explain the difference between analog and binary input devices.

2. Why does a low-voltage AC control system give better control than a line voltage control system?

3. Why does a furnace blower control keep the blower ON after the thermostat setpoint is reached?

4. Why should the evaporator motor in an air conditioning system be on a separate control from the compressor and condenser motors?

5. Why is an economizer damper and control package used?

16 Control Systems

Name _____   Date _____

Activity — Identification

_____ **1.** High-pressure limit control

_____ **2.** Fuse

_____ **3.** Compressor motor

_____ **4.** Disconnect

_____ **5.** Thermostat

_____ **6.** Transformer

_____ **7.** Compressor motor relay coil

_____ **8.** Evaporator blower relay

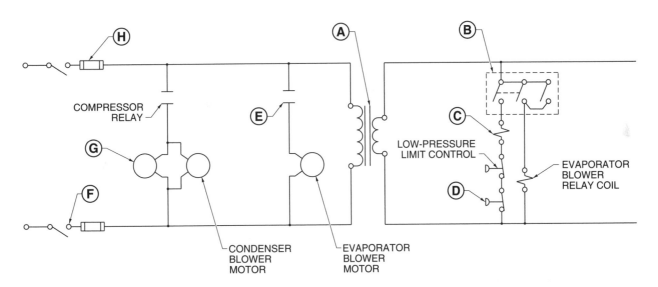

Activity — Power Controls

Equip a refrigeration unit or trainer with a thermostatic expansion device. Turn the unit ON by turning the thermostat setting below the room temperature. Turn the unit OFF by opening the disconnect switch. Turn the unit ON by closing the disconnect switch. Turn the unit OFF, remove a fuse. Replace the fuse and turn the unit ON.

_____ **1.** _____ controls control the operation of the unit.

_____ **2.** Which side of the fuse is removed first?

Activity — Control Functions

1. List the controls that are considered safety controls on a typical forced-air heating system.

2. Which controls function when the furnace is operating?

3. List the controls that are considered safety controls on an air conditioning system.

4. Which controls function when the blower is operating constantly but the unit is not running?

Activity — Operating Controls

Equip a refrigeration unit or trainer with a thermostatic expansion device. Turn the unit ON by turning the thermostat setting below the room temperature. Turn the unit OFF by turning the thermostat to a higher setting than the room temperature. Turn the unit ON by turning the thermostat to a lower setting than the room temperature. Turn the unit OFF, disconnect the wires from the secondary side of the transformer. Replace the wires to appropriate terminals and turn the unit ON.

1. What type of controls control the operation of the unit?

2. If wires are reconnected opposite, what is the result?

Activity — Safety Controls

Equip a refrigeration unit or trainer with a thermostatic expansion device. Turn the unit ON by turning the thermostat setting below the room temperature. Turn the unit OFF by turning the pressure setting on the low-pressure switch to a higher pressure than the operating low pressure of the unit. Turn the unit ON by turning the pressure setting on the low-pressure switch to the normal operating low pressure of the unit. Turn the unit OFF by turning the pressure setting on the high-pressure switch to a lower setting than the operating high pressure of the unit. Turn the unit ON by turning the high-pressure setting on the high-pressure switch to the normal operating pressure of the unit.

1. What type of controls control the operation of the unit?

2. Which method of turning the unit OFF and ON is the safest? Why?

Activity — Inspection

After obtaining permission from the proper authority, inspect the heating and cooling systems in a commercial building. Complete the inspection report.

Date _____   Name of building _____

Owner/manager_____   Address _____

Type of heating system _____   Type of cooling system _____

1. Identify the type of control system used in the building.

2. List the controls used for heating the building.

3. List the controls used for cooling the building.

4. How is the humidity controlled in the building?

5. List the controls used for ventilating the building.

6. Are outside air dampers used on the system?

7. Is an automatic control system used on the dampers?

16 Control Systems

Name _____ Date _____

True-False

T F **1.** Operating controls cycle equipment ON or OFF.

T F **2.** A universal input-output controller is designed to control only one type of HVAC system.

T F **3.** A building automation system output device changes the state of a controlled device in response to a command from the controller.

T F **4.** A stack switch senses flue gas temperatures.

T F **5.** A thermistor is a sensing element that changes electrical resistance in response to change in pressure.

Completion

_____ **1.** A pneumatic control system uses _____ as the control signal.

_____ **2.** A(n) _____ is a device that surrounds a wire and detects the electromagnetic field due to electricity passing through the wire.

_____ **3.** A(n) _____ input device produces an ON or OFF signal.

_____ **4.** Most residential control circuit transformers are _____ transformers with a secondary voltage of 24 V.

_____ **5.** Relays use a coil to create magnetism to open or close one or more sets of _____.

Multiple Choice

_____ **1.** A flame _____ is an electronic combustion safety control that uses the flame as an electrical conductor.
   A. rod
   B. surveillance control
   C. burner control
   D. temperature sensor

_____ **2.** A bellows element is an accordion-like device that converts _____ variation into mechanical movement.
   A. pressure
   B. temperature
   C. humidity
   D. air flow

_____ **3.** An economizer package is the damper and controls that bring in _____ air to cool building spaces.

    A. recirculated   C. dehumidified

    B. outdoor   D. air-conditioned

_____ **4.** Analog input is a device that senses a variable such as temperature, pressure, or humidity, and causes a proportional electrical signal change at the _____.

    A. dampers   C. gas valve

    B. dehumidifier   D. controller

_____ **5.** A(n) _____ diagram uses lines and symbols to represent the electrical circuits and components in an electrical control system.

    A. wiring   C. schematic

    B. plan view   D. all of the above

Short Answer

1. Why must the condenser blower motor run when the compressor runs?

2. What is the function of a humidistat in a control system?

3. Explain the function of a sensor in a control system.

4. What operating characteristic of a relay makes it useful in a control circuit?

Name _____ Date _____

True-False

T   F   **1.** Heating and cooling loads exist because of a temperature difference between the indoor and outdoor temperatures, infiltration or ventilation in the building, and internal loads.

T   F   **2.** The winter dry bulb temperature for a building is the coldest temperature expected in January for the location.

T   F   **3.** Exposed surfaces are building surfaces that are exposed to outdoor temperatures.

T   F   **4.** Any device that produces heat inside a building must be compensated for by a cooling load.

T   F   **5.** A conduction factor for a building component has time factored into it.

T   F   **6.** Infiltration occurs when air flows through a furnace filter.

T   F   **7.** Heat gain from building occupants varies depending on the activity of the occupants.

T   F   **8.** Equivalent temperature difference is the design temperature difference, which is adjusted for heat gain from appliances.

T   F   **9.** Estimating the number of occupants for residential and commercial buildings is calculated by the same method.

T   F   **10.** Net wall area is the area of a wall after the area of windows, doors, and other openings have been subtracted from gross wall area.

Completion

_____ **1.** Factors are numerical values that represent the _____ produced or transferred under some specific condition.

_____ **2.** _____ are data that are unique to a building relating to the specific location of the building and the specifications of the particular building.

_____ **3.** A building _____ is a main part of a building structure such as the exterior walls.

_____ **4.** _____ is the process that occurs when outdoor air is brought into a building.

_____ **5.** _____ gain is heat gain caused by radiant energy from the sun that strikes opaque objects.

_____ 6. _____ forms are documents that are used by design technicians for arranging the heating and cooling load variables and factors.

_____ 7. A(n) _____ form is a blank table that is divided into columns and rows by vertical and horizontal lines.

_____ 8. A(n) _____ factor is a factor that represents the amount of heat that flows through a building component because of a temperature difference.

_____ 9. A(n) _____ form is a preprinted form consisting of columns and rows that identify required information.

_____ 10. _____ is the geographic direction a wall faces.

Short Answer

1. Why must the heating and cooling loads of a building be known?

2. Define indoor design temperature.

3. Define temperature swing.

4. Explain design temperature difference.

5. Define outdoor design temperature.

6. Define air change factor.

7. Why does a cooling design temperature disregard the most extreme temperatures that occur?

8. Within what temperature range should the indoor temperature selected for load calculations usually fall?

Heating and Cooling Loads

Activities

Name _____   Date _____

Activity — Design Temperature

Answer the questions based on the outdoor design temperature table.

_____ **1.** What is the winter dry bulb temperature for Memphis, Tennessee?

_____ **2.** The summer wet bulb temperature for Houston, Texas is _____ °F.

_____ **3.** What is the summer dry bulb temperature for Dallas, Texas?

OUTDOOR DESIGN TEMPERATURE					
State and City	Lat.§	Winter	Summer		
		DB*	DB*	Daily Range*	WB*
TENNESSEE					
Chattanooga	35	13	96	22	78
Knoxville	35	13	94	21	77
Memphis	35	13	98	21	80
Nashville	36	9	97	21	78
TEXAS					
Austin	30	24	100	22	78
Corpus Christi	27	31	95	19	80
Dallas	32	18	102	20	78
El Paso	31	20	100	27	69
Fort Worth	32	17	101	22	78
Galveston	29	31	90	10	81
Houston	29	27	96	18	80

*in °F   §in degrees

Activity — Area of Exposed Surfaces

_____ **1.** What is the gross wall area?

_____ **2.** The area of the door is _____ sq ft.

_____ **3.** The area of the two windows is _____ sq ft.

_____ **4.** The area of all openings is _____ sq ft.

_____ **5.** The net wall area is _____ sq ft.

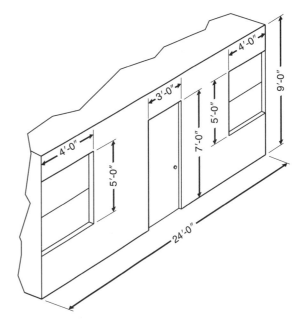

Activity — Conduction Factors

Obtain a 1′ square cardboard box with a lid. Place the box on a stand so all sides are exposed to room air. Suspend a low wattage light bulb in the center of the box. Different wattage bulbs may be used to get the desired results. Place a remote thermostat lead in the box and a remote thermostat lead outside the box. Connect the light bulb cord to a wattmeter. Connect the wattmeter to an electrical outlet. Place the lid on the box and seal it with tape. Turn the light ON.

Time (in minutes)	Temperature (in °F)
Initial	
1	
2	
3	
4	
5	
6	
7	
8	
9	
10	

1. Record the temperature of the air inside and outside the box at intervals of approximately 1 minute until the temperature inside the box stabilizes.

Calculate the amount of heat that passes through the cardboard using the surface area of the box, a conduction factor of .48 for cardboard, and the difference between the final temperature inside the box and the temperature outside the box.

_____ 2. The conduction rate of the cardboard box is _____ Btu/hr/sq ft.

3. If some other material is substituted for one side of the box, could the conduction factor of the new material be calculated? Explain.

4. Why must the temperature be stabilized while calculating the conduction rate?

Name _____   Date _____

Calculations

_____ **1.** If a building is 63′ long, 34′ wide, and has a 12′ high ceiling, what is the volume of the building?

_____ **2.** The wall of a building is 47′ wide and 27′ high. The gross wall area is _____ sq ft.

_____ **3.** If a building has 27 occupants and 30 cfm of air per person is required for ventilation, what is the volumetric flow rate of ventilation air?

_____ **4.** If 1200 cfm of ventilation air is required for a building and the design temperature difference is 42°F, what is the heat loss due to ventilation air?

_____ **5.** If the gross wall area of one wall of a building is 360 sq ft and the area of window and door openings is 41 sq ft, what is the net wall area?

_____ **6.** If a building is 100′ long, 50′ wide, and 9′ high, what is the volume of the building?

_____ **7.** A wall of a building is 25′ wide and 8′ high. The wall contains one window 6′ wide and 4′ high. The net wall area of the wall is _____ sq ft.

_____ **8.** If a building has 19 occupants and 30 cfm of air per person is required for ventilation, the volumetric flow rate of ventilation air is _____ cfm.

Short Answer

1. How is the number of occupants estimated for load calculations for a residential building?

2. Where are design temperatures required for calculating loads for a given building found?

3. How is the area of a sloping ceiling calculated for heating and cooling loads?

4. Why must solar gain be calculated for cooling loads?

5. Why are factors entered based on exposure for cooling load calculations?

6. Why are areas calculated separately for each building component?

7. What is the advantage of using a columnar load form?

8. State the law of thermodynamics that accounts for the transfer of heat through a building component.

9. Define heat transfer factor.

10. Define design temperature.

Name _____ Date _____

Short Answer

1. Define computer load calculation program.

2. Why is data for calculating cooling loads usually entered on a load form first?

3. How is the number of people that occupy a residential building determined?

4. What is the best source of information for dimensions of components of a building that is already built?

5. Why must adjustments be made to the distribution system of a building after it is finished?

True-False

T   F   **1.** A heat transfer factor is the same as a conduction factor.

T   F   **2.** Total heat loss or gain for a building is the sum of all the room or zone subtotals.

T   F   **3.** Duct loss and duct gain adjustments are made regardless of where in a building the ducts run.

T   F   **4.** When analyzing load calculations, if the total loads from a computer load calculation program are close to a standard, no other comparisons are required.

T F **5.** Most commercial computer load calculation programs are written so a printout of the results show subtotals of losses or gains for the rooms or zones in a building.

T F **6.** The results of a computer load calculation program are assumed to be correct.

T F **7.** All computer load calculation programs include all variables and factors in the initial calculations and require no adjustments.

T F **8.** Some difference usually exists between computer-aided load figures and conventional load figures because different factors are used for each load calculation method.

T F **9.** A winter indoor design temperature of 70°F is usually selected to provide comfort for a normally clothed person during winter.

T F **10.** Plans are drawings of a building that show dimensions, construction materials, location, and arrangement of the spaces within a building.

Completion

_____ **1.** For residential cooling applications, _____ Btu/hr of heat gain is used for appliances and cooking processes.

_____ **2.** For residential cooling applications, heat gain from people is _____ Btu/hr of sensible heat per person with a latent heat allowance added.

_____ **3.** Variables and _____ are entered on a load form and are used to calculate heat losses and gains.

_____ **4.** A summer indoor design temperature of _____°F is usually selected to provide comfort for a normally clothed person during summer.

_____ **5.** _____ are written supplements to plans that describe the materials used for a building.

_____ **6.** Heating and cooling loads are calculated by a conventional method or with a _____.

_____ **7.** _____ is the opening around windows, doors, or other openings in a building such as between the foundation and framework of a building.

_____ **8.** Factors used in calculating heating and cooling loads include conduction factors, solar gain, _____ factors, and factors for heat produced by people, lights, and appliances.

_____ **9.** _____ components are the main part of a building structure such as the exterior walls, windows, doors, ceiling, and floor.

_____ **10.** To oversize heating and air conditioning equipment components, an equipment _____ factor is used.

Name _____ Date _____

Activity — Job Information

Use the prepared load form on page 201–202 for all of the activities in this section. Select a residential building for a heating and cooling load calculation. Enter the job information on the top of the load form.

1. List the data normally entered on the top of a load form.

Activity — Design Temperature

Select a winter indoor design temperature for the building. Enter the temperature on the load form. Refer to the outdoor design temperature tables on page 192–195. Find the winter outdoor design temperature for the area. Enter the temperature on the load form. Subtract the winter outdoor design temperature from the winter indoor design temperature to obtain the winter design temperature difference. Enter the temperature on the load form.

Find the summer outdoor design temperature for the area. Enter the temperature on the load form. Select a summer indoor design temperature for the building. Enter the temperature on the load form. Subtract the summer indoor design temperature from the summer outdoor design temperature to obtain the summer design temperature difference. Enter the temperature on the load form.

1. How is the outdoor design temperature found for an area that is not covered in an outdoor design temperature table?

2. Is the wet bulb temperature difference used in calculating heating loads?

Activity — Variables

On a separate sheet of paper, draw a sketch showing all dimensions needed for calculating heating and cooling loads. List all details required for selecting factors. Enter the name of the rooms, running feet of exposed walls, dimensions, ceiling height, and exposure of each room in the building on the load form. Calculate the gross exposed wall area, area of windows and glass doors for heating, and areas of windows and glass doors based on exposure for cooling for each room in the building. Enter the data on the load form.

1. Is the gross exposed wall area used for calculating losses or gains?

2. What is the difference between area and volume?

3. Why must the area of windows and glass doors be entered based on their exposure for a cooling load calculation?

Calculate the area of other doors, net exposed wall area, ceilings, and floors for each room in the building. Enter the data on the load form.

4. How is the area of the ceiling calculated if a building has a sloped ceiling?

5. How is the area of the floor calculated for a split-level residence?

6. Are the dimensions for ceiling and floor areas taken from the inside or outside dimensions?

Determine the number of people occupying the building. Enter the data on the load form. Enter the heat gain for appliances and cooking processes on the load form.

7. Why is heat gain from lights not included as an additional factor when calculating a residential cooling load?

8. What is included in the kitchen allowance when calculating a residential cooling load?

Activity — Factors

Refer to the heat transfer factor tables on page 188–191. Determine the heat transfer factors for the windows and glass doors, other doors, net exposed walls, ceilings, and floors. Enter the factors on the load form. Determine the cooling heat transfer factors for the windows and glass doors based on the exposure, other doors, net exposed walls, ceilings, and floors. Enter the factors on the load form.

1. How are infiltration factors accounted for?

2. Factors for heat transfer are based on what two variables?

Activity — Calculations

Multiply the heating factors by the building component areas to obtain the individual building component heating loads. Add the heating values horizontally to obtain the building component heat loss subtotals. Add the values vertically to obtain the subtotal heat loss for the individual rooms.

Multiply the cooling factors by the building component areas to obtain the individual building component cooling loads. Add the cooling values horizontally to obtain the building component heat gain subtotals. Add the values vertically to obtain the sensible heat gain for the individual rooms. Multiply the sensible heat gain by 1.3.

1. Why is the sensible heat gain multiplied by a factor of 1.3?

2. Is the total heat gain found from the load form the figure that is used for designing the distribution system? Explain.

Activity — Equipment Adjustment

Obtain manufacturer's specifications sheets for furnaces and air conditioners. Select a furnace and air conditioner for the building. Calculate the equipment ratio.

1. Why is it necessary to make an adjustment for equipment size?

2. If it is necessary to do so, when is duct loss or gain added to a load?

Name _____ Date _____

Completion

Contractor: _Best Heating & Cooling_

Name of Job: _Johnson Residence_ Date _4/9_

Address: _1841 N. Adams_ By: _J.G._

Winter: Indoor Design Temp. _78°F_ Outdoor Design Temp. _17°F_ Design Temperature Difference _58°F_

Summer: Outdoor Design Temp. _84°F_ Indoor Design Temp. _75°F_ Design Temperature Difference _9°F_

					1 LR			4 BR3			5 MAIN BR			Building
1	Name of Room													Component
2	Running Feet of Exposed Wall				42			25			14			Subtotals
3	Room Dimensions				28' × 14'			12' × 13'			14' × 14'			
4	Ceiling Height	Exposure			8'	S - W		8'	E - S		8'	S		
	Types		Factors		Area	Btu/hr	Btu/hr	Area	Btu/hr	Btu/hr	Area	Btu/hr	Btu/hr	Btu/hr	Btu/hr
	Exposure		H	C		H	C		H	C		H	C	H	C
5	Gross	N													
	Exposed	S			224			96			112				
	Wall	E						104							
	Area	W			112										
6	Windows and Glass Doors (H)	60			40	2400		15	900		15	900			
7	Windows and Glass Doors (C)	N	14												
		E&W	44												
		S	23		40		920	15		345	15		345		
8	Other Doors	145	8.6		21	3045	181								
9	Net Exposed Walls	4	1.7		275	1100	468	185	740	315	97	388	165		
10	Ceilings	2	1		392	784	392	156	312	156	196	392	196		
11	Floors	4	.8		392	1568	314	156	624	125	196	784	157		
12	People @ 300 Appliances @ 1200				2P		600								
														Totals	

Refer to partial Load Form on page 131.

_____ 1. The dimensions of the living room are _____.

_____ 2. The net exposed wall area for bedroom 3 is _____ sq ft.

_____ **3.** The heat loss through the windows and glass doors in the living room is _____ Btu/hr.

_____ **4.** The exposure of the walls in the living room are south and _____.

_____ **5.** The heating heat transfer factor for the floors of the building is _____.

_____ **6.** The cooling heat transfer factor for the windows and glass doors on the north exposure of the building is _____.

_____ **7.** The heat loss through the ceiling of bedroom 3 is _____ Btu/hr.

_____ **8.** The floor area of the main bedroom is _____ sq ft.

_____ **9.** The height of the ceiling is _____'.

_____ **10.** The cooling heat transfer factor for the walls of the building is _____.

Calculations

_____ **1.** How many people are entered on a load calculation form for a residential building with four bedrooms?

_____ **2.** If a winter indoor design temperature of 70°F is used for an application and the outdoor design temperature is 13°F, the design temperature difference is _____°F.

_____ **3.** If the wall of a building is 42' long and 9' high, what is the gross wall area?

_____ **4.** A wall has a gross wall area of 243 sq ft. Two windows in the wall are 6' × 4'. The net wall area of the wall is _____ sq ft.

_____ **5.** Outdoor air at a dry bulb temperature of 55°F and a flow rate of 200 cfm is introduced into a building for ventilation. The dry bulb temperature of the indoor air is 75°F. The heat loss of the air is _____ Btu/hr.

_____ **6.** A building has a total heat loss of 63,027 Btu/hr. The ductwork in the building is located in an unheated crawlspace. The duct loss multiplier is 1.10. The adjusted heat loss due to ductwork is _____ Btu/hr.

_____ **7.** Find the cooling equipment ratio if the cooling load for a building is 34,000 Btu/hr and a 38,000 Btu/hr air conditioner is chosen.

_____ **8.** Outdoor air at a dry bulb temperature of 40°F and a flow rate of 150 cfm is introduced into a building for ventilation. The dry bulb temperature of the indoor air is 75°F. The heat loss of the air is _____ Btu/hr.

_____ **9.** A building has a total heat loss of 42,753 Btu/hr. The ductwork in the building is located in an unheated crawlspace. The duct loss multiplier is 1.30. The adjusted heat loss due to ductwork is _____ Btu/hr.

_____ **10.** The total heating load for a building is 86,500 Btu/hr. A furnace with an output of 100,000 Btu/hr is selected. The heating equipment ratio is _____.

Name _____   Date _____

Completion

_____ 1. Availability and _____ are two factors considered when deciding what fuel or energy to use for heating a building.

_____ 2. _____ furnaces are used where the supply ductwork is located above the furnace.

_____ 3. Furnaces with output capacities up to _____ Btu/hr are generally used for residential applications.

_____ 4. _____ rating is the amount of heat produced per unit of fuel.

_____ 5. The _____ is the central component in a forced-air heating system.

_____ 6. A design _____ is a person who has the knowledge and skill to plan heating and/or air conditioning systems.

_____ 7. Catalogs and engineering _____ sheets are the best source of information about heating and air conditioning equipment.

_____ 8. _____ rating is the amount of heat a furnace produces in 1 hour.

_____ 9. For a typical application, the air flow rate in a building space should be less than _____ cfm per square foot of floor space if the area is occupied by people.

_____ 10. The air flow rate required for heating a building is a function of the furnace output rating and the _____ increase of the air through the heat exchanger.

Short Answer

1. What is the main purpose of a terminal device in a hydronic heating system?

2. What is the main function of a circulating pump in a hydronic system?

3. Why must the output rating of a furnace be equal to or greater than the heat loss of the building it heats?

4. Define design conditions.

5. Where does a heat reclaim system obtain heat?

True-False

T   F   **1.** Combustion furnaces usually operate with a temperature increase of approximately 80°F to 100°F.

T   F   **2.** The central component in a forced air conditioning system is the air distribution system.

T   F   **3.** Package air conditioning systems are used when the evaporator is located in one part of the air distribution system, and the condenser and compressor are located in another part of the air distribution system.

T   F   **4.** The cooling capacity of an air conditioner must be as great or greater than the calculated cooling load of the building.

T   F   **5.** When using a temperature swing table to select an air conditioner, a larger air conditioner must be used to offset the same heat gain to the building.

T   F   **6.** The air flow rate across the evaporator coil of an air conditioner is important for the unit to operate properly.

T   F   **7.** Most air conditioners operate with approximately 400 cfm of air across the evaporator coil per ton of cooling.

T   F   **8.** Piping systems are categorized by the arrangement of the piping loop that carries the water.

T   F   **9.** The temperature control required in a building determines the terminal device to be used.

T   F   **10.** Absorption chillers are used in buildings of all sizes because the chillers are compact and can be adapted to all conditions.

Name _____   Date _____

Activity — Identification

_____   **1.** Distribution system

_____   **2.** Blower

_____   **3.** Grill

_____   **4.** Controls

_____   **5.** Register

_____   **6.** Furnace

Activity — Fuel Supply

Contact local fuel suppliers and utilities to determine the cost of the most common fuels and energies used for heating and cooling. Record the values on form 1. Determine the cost of the fuel or energy per 100,000 Btu/hr (therm). Record the values on form 1. Rate the fuels according to cost. Assign the number 1 to the least expensive fuel and assign the number 5 to the most expensive fuel.

Determine the relative availability of each fuel. Assign the number 1 to the most available fuel and assign the number 5 to the least available fuel. Add the numbers across the form to determine which fuel or energy is most practical based on cost and availability. The lowest number indicates the best fuel to use.

Fuel or energy	Unit cost	Cost/therm	Cost rating	Availability rating	Total
Natural gas					
LP gas					
Fuel oil					
Coal					
Electricity					

FORM 1

1. Which fuel is the most practical based on cost and availability?

2. What other factors may be considered when selecting a fuel?

Activity — Forced-Air Distribution Systems

After obtaining permission from an authority at each site, survey a residential building and a commercial building that contain forced-air distribution systems. Record the site, address, person in charge, person who grants permission, distribution system location, and equipment location.

Residential site _____   Address _____

Person in charge _____   Permission granted by _____

Distribution system location

Equipment location

Commercial site _____   Address _____

Person in charge _____   Permission granted by _____

Distribution system location

Equipment location

1. List the factors other than ductwork location that determine where heating and air conditioning equipment must be located.

Activity — Boilers

After obtaining permission from an authority at each site, survey a residential building and a commercial building that contain electric and combustion boilers. Record the site, address, person in charge, person who grants permission, boiler, and why it was chosen.

Residential site _____   Address _____

Person in charge _____   Permission granted by _____

Boiler

Why chosen

Commercial site _____   Address _____

Person in charge _____   Permission granted by _____

Boiler

Why chosen

1. List the factors that determine what type of boiler to use for a particular application.

Activity — Hydronic Distribution Systems

After obtaining permission from an authority at each site, survey a residential building and a commercial building that contain hydronic distribution systems. Record the site, address, person in charge, person who grants permission, distribution system, and why it was chosen.

Residential site _____   Address _____

Person in charge _____   Permission granted by _____

Distribution system

Why chosen

Commercial site   _____   Address _____

Person in charge   _____   Permission granted by _____

Distribution system

Why chosen

On a separate sheet of paper, draw a schematic diagram of one of the distribution systems showing the location of the major parts and the direction of the water flow in the system.

1. Why was each system selected for the building?

2. Would another system be more appropriate for each application? Explain.

Name _____   Date _____

Calculations

_____   1. A furnace has an output rating of 160,000 Btu/hr and a temperature increase of 90°F. The air flow rate is _____ cfm.

_____   2. A room has an area of 400 sq ft. A furnace blower circulates 800 cfm of air. Does the air flow rate create a draft?

_____   3. An air conditioner has a cooling capacity of 58,000 Btu/hr. The nominal size of the air conditioner is _____ tons.

_____   4. An air conditioner has a cooling capacity of 37,000 Btu/hr. The required air flow rate across the evaporator coil is _____ cfm.

_____   5. What is the water flow rate through a hydronic piping system if the boiler output rating is 246,000 Btu/hr and the system has a water temperature drop of 17.6°F?

_____   6. A furnace has an output rating of 210,000 Btu/hr and a temperature increase of 85°F. The air flow rate is _____ cfm.

_____   7. A refrigeration system has a cooling capacity of 251,000 Btu/hr. The nominal size of the refrigeration system is _____ tons.

_____   8. An air conditioner has a cooling capacity of 118,500 Btu/hr. The required air flow rate across the evaporator coil is _____ cfm.

_____   9. Air enters the evaporator coil of an air conditioner at 80°F db and 70°F wb. The air conditioner has a cooling capacity of 52,600 Btu/hr and an air flow rate of 1670 cfm. The air leaves the evaporator coil at 85% rh and _____°F db.

_____   10. A chiller has a cooling capacity of 465,700 Btu/hr. The total temperature increase through the terminal devices in the system is 22.4°F. The required water flow rate is _____ gpm.

_____   11. A natural gas-fired furnace with a conventional burner and heat exchanger has an input rating of 300,000 Btu/hr. How much heat is produced?

_____   12. What is the output rating of an electric furnace that has an input rating of 45,000 Btu/hr?

Short Answer

1. What two factors must be considered when determining the air flow rate for a heating application?

2. Why are direct-fired heaters used almost exclusively for industrial applications?

3. What is the main reason for using a hydronic heating system when the spaces to be heated are located a distance from the heat source?

4. What two factors should be considered when selecting a piping system for a building?

5. What two factors must be considered when sizing a circulating pump for a hydronic heating system?

6. What two factors must be considered when choosing a chiller for an application?

7. What two factors must be considered when calculating the water flow rate for a hydronic air conditioning system?

8. What kind of hydronic application requires a three-pipe piping system?

Name _____ Date _____

True-False

T	F	**1.** A manometer is a device that measures liquid pressure.
T	F	**2.** Water column is the pressure exerted by a square foot of a column of water.
T	F	**3.** Static pressure drop is the decrease in air pressure caused by friction between the air moving through a duct and the internal surfaces of a duct.
T	F	**4.** An equal friction chart shows the relationships between the air flow rate, static pressure drop, duct size, and air velocity.
T	F	**5.** Friction loss is the increase in air pressure due to the friction of the air moving through a duct.
T	F	**6.** Total static pressure drop in a distribution system is the sum of the static pressure drop through all duct sections, fittings, transitions, and accessories that add resistance to air flow in the distribution system.
T	F	**7.** Velocity pressure is the air pressure in a duct that is measured parallel to the direction of air flow.
T	F	**8.** Fitting loss coefficients are values that represent the ratio between the total static pressure loss through a fitting and the dynamic pressure at the fitting.
T	F	**9.** A duct is sized to allow the required amount of air flow for heating and cooling building spaces.
T	F	**10.** Large, sophisticated distribution systems are sized using the velocity reduction method.
T	F	**11.** The equal friction method of duct sizing considers that the static pressure is approximately equal at each branch takeoff in the distribution system.
T	F	**12.** Small distribution systems are designed for 1″ WC to 6″ WC static pressure drop per 100′ of duct.
T	F	**13.** To function most efficiently, the aspect ratio of a rectangular duct should not be more than 3 to 4 times the height.
T	F	**14.** The equal velocity method of duct sizing considers that the air velocity decreases at each branch takeoff.
T	F	**15.** The air flow pattern from a register is controlled by the shape of the ductwork.

Completion

_____ 1. Air _____ is the speed at which air moves from one point to another.

_____ 2. _____ pressure is air pressure in a duct measured at right angles to the direction of air flow.

_____ 3. _____ pressure drop is the pressure drop in a duct fitting or transition caused by air turbulence as the air flows through the fitting or transition.

_____ 4. _____ ratio is the ratio between the height and width of a rectangular duct.

_____ 5. _____ area of a register is the face area of the register minus the area blocked by the frame or vanes.

_____ 6. _____ static pressure drop is the pressure drop per unit length of duct for a given size of duct at a given air flow rate.

_____ 7. _____ pressure drop is the total static pressure drop in a forced-air distribution system.

_____ 8. _____ drawings are drawings of the floor plan of a building that show walls, partitions, windows, doors, fixtures, and other details that affect the location of the ductwork, registers, and grills.

Short Answer

1. List the three most common duct systems used in heating and air conditioning distribution systems.

2. What causes friction loss in a duct system?

3. What two variables must be considered when sizing a blower?

4. Describe the static regain method of duct sizing.

5. How is the static pressure drop through an individual duct section found?

Name _____   Date _____

Activity — Duct Air Flow

Set up a section of ductwork by connecting 10′ to 30′ of round or rectangular duct to a blower outlet. Make sure the ductwork has a greater capacity than the output of the blower. Measure and record the inside dimensions of the duct.

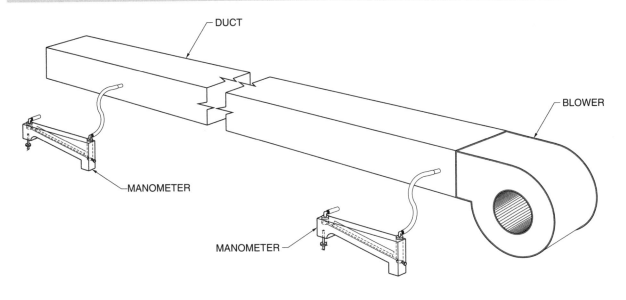

_____   **1.** The duct dimensions are _____.

Turn the blower ON. Measure and record the velocity of the air in the duct with a manometer. Refer to Friction Loss Chart on page 196.

_____   **2.** The velocity of the air is _____ fpm.

Calculate and record the volumetric flow rate of air flowing through the duct from the duct dimensions and the velocity of the air.

_____   **3.** The air flow rate is _____ fpm.

4. What is the relationship between duct size and air velocity?

5. If a smaller duct is used, would the velocity be lower? Explain.

6. If a blower that delivered more air is used with the same size duct, would the velocity be greater? Explain.

Activity — Static Pressure Drop

Set up a section of ductwork by connecting 10′ to 30′ of round or rectangular duct to a blower outlet. Make sure the ductwork has a greater capacity than the output of the blower. Turn the blower ON. Measure and record the static pressure at both ends of the duct.

_____ **1.** The duct static pressure drop at the beginning of the duct is _____″ WC.

_____ **2.** The duct static pressure drop at the end of the duct is _____″ WC.

From the two readings, calculate the total static pressure drop for the duct section.

_____ **3.** The total static pressure drop for the duct section is _____″ WC.

Calculate the total static pressure drop for 100′ of duct.

_____ **4.** The total static pressure drop for 100′ of duct is _____″ WC.

5. Explain the relationship between static pressure at a given point in a duct and air movement at that point.

20 Forced-Air System Design

Name _____ Date _____

Calculations

_____ **1.** The velocity of the air entering a fitting is 1350 fpm. The velocity pressure in the fitting is _____ ″ WC.

_____ **2.** A grill in a forced-air distribution system must handle 1200 cfm of air at a face velocity of 500 fpm. The required free area is _____ sq in.

_____ **3.** A building has a heating load of 56,000 Btu/hr. The temperature increase through the furnace is 84°F. Find the air flow rate required to provide comfort in the building.

_____ **4.** A room in a building has a cooling load of 7500 Btu/hr. The air flow rate required to provide comfort in the room is _____ cfm.

_____ **5.** The blower in a forced-air distribution system moves 1100 cfm of air through a 2′ × 1′ duct. The velocity of the air is _____ fpm.

_____ **6.** A grill in a forced-air distribution system must handle 1800 cfm of air at a face velocity of 700 fpm. The required free area is _____ sq in.

_____ **7.** A room in a building has a heating load of 6700 Btu/hr. The temperature increase through the furnace is 90°F. Find the air flow rate required to provide comfort in the room.

_____ **8.** A duct has a design static pressure drop of .08″ WC. The duct section is 160′ long. The static pressure drop of the duct section is _____ ″ WC.

_____ **9.** The velocity of the air entering a fitting is 1100 fpm. The velocity pressure in the fitting is _____ ″ WC.

_____ **10.** A duct in an air distribution system is 46′ long. The equivalent length of a fitting in the duct is 15′. The duct is sized for .125″ WC per 100′ of duct. What is the static pressure drop in the duct?

_____ **11.** A grill in a distribution system must handle 1500 cfm of air at a face velocity of 500 fpm. The height of the duct must be 18″. The width of the grill is _____ ″.

_____ **12.** A duct system has a total equivalent length of 365′. The duct has a design static pressure drop of .085″ WC per 100′ of duct. The total static pressure drop is _____ ″ WC.

Short Answer

1. List the four variables shown on an equal friction chart.

2. Describe the velocity reduction method of duct sizing.

3. Describe a U-tube manometer.

4. The size of a register is determined by what two factors?

5. What is the disadvantage of sizing duct by the static regain method.

Completion

Refer to Friction Loss Chart on page 196.

_____ 1. The air flow rate in a section of duct is 10,000 cfm. The static pressure drop is 1.0″ WC per 100′ of duct. The diameter of the duct is _____″.

_____ 2. What is the diameter of a duct that carries 1000 cfm of air at a static pressure drop of .1″ WC per 100′ of duct?

_____ 3. The velocity of the air flowing through a 12″ diameter duct is 1200 fpm. The flow rate of the air is _____ cfm.

_____ 4. What is the diameter of a duct that carries 200 cfm of air at a static pressure of .1″ WC per 100′ of duct?

_____ 5. What is the velocity of the air in a 9″ diameter duct with 1200 cfm of air flowing through it?

Name _____   Date _____

True-False

T   F   **1.** The rate of water flow through a pipe or orifice is a function of the pressure exerted on the water and the size of the pipe.

T   F   **2.** Water pressure can be expressed in feet of head and pounds per square inch.

T   F   **3.** A pipe section is a length of pipe that runs from one fitting to the next fitting.

T   F   **4.** If a piping system is sized for heating and cooling, the smaller flow rate is used for sizing the components of the system.

T   F   **5.** Volumetric flow rate of water is measured in feet per minute.

T   F   **6.** Friction head is the effect of gravity in a pipe.

T   F   **7.** After the piping system is chosen, the terminal devices for each building space are chosen based on the heating and cooling requirements.

T   F   **8.** Copper tubing is used for large applications, and iron or steel pipe is used for small applications.

T   F   **9.** The procedure for sizing a piping system depends on the type of piping system used.

T   F   **10.** The pipe size remains the same throughout a one-pipe series system because the flow rate remains the same.

Completion

_____   **1.** _____ head is the sum of friction head and static head.

_____   **2.** _____ head is the weight of water in a column above a datum line.

_____   **3.** A(n) _____ is a unit of measure equal to $\frac{1}{1000}''$.

_____   **4.** A(n) _____ senses water temperature in a boiler or chiller and operates the equipment to maintain the setpoint temperature.

_____   **5.** In a straight pipe section, friction is a function of the size of the pipe, amount of water flow, and the _____ of the surface of the pipe.

_____   **6.** For most small- to medium-size applications, a design static pressure drop of about _____′ of head is used.

_____   **7.** A piping system consists of iron or steel pipe or copper tubing, _____, and valves.

_____ 8. _____ drawings are drawings of the floor plan of a building that show walls, partitions, windows, doors, fixtures, and other details that affect the location of the piping and terminal devices.

_____ 9. In a two-pipe _____-return system, the supply pipe becomes smaller as it goes out from the boiler or chiller, and the return pipe becomes larger as it comes back to the boiler or chiller.

_____ 10. For a typical small- to medium-size application, a temperature difference of about _____°F may be used for both heating and cooling applications.

Short Answer

1. Define feet of head.

2. How is the pressure drop through a pipe fitting found?

3. Describe an outdoor reset control.

4. How is room temperature controlled in a hydronic system?

5. Circulating pumps are selected based on what two factors?

Name _____   Date _____

Activity — Temperature Difference

Determine the temperature difference through the terminal devices in hydronic system 1.

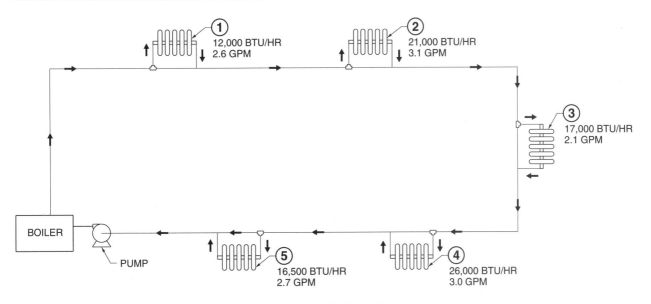

HYDRONIC SYSTEM 1

_____ **1.** The temperature difference through terminal device 1 is _____°F.

_____ **2.** The temperature difference through terminal device 2 is _____°F.

_____ **3.** The temperature difference through terminal device 3 is _____°F.

_____ **4.** The temperature difference through terminal device 4 is _____°F.

_____ **5.** The temperature difference through terminal device 5 is _____°F.

6. In what type of piping system is the temperature difference across a terminal device directly proportional to the heating load on the terminal device? Explain.

7. What does the temperature difference through each pipe section and terminal device represent?

Activity — Volumetric Flow Rate

In hydronic system 2, use the heating load on the terminal devices at the temperature drops shown to determine the volumetric flow rate of water through each terminal device.

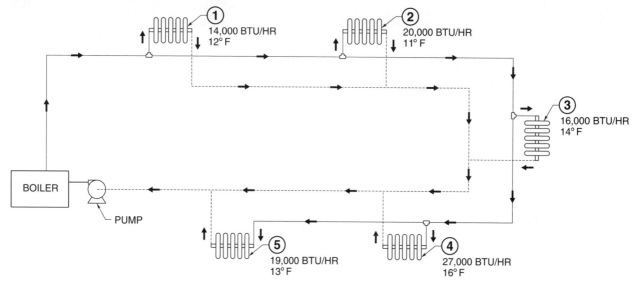

HYDRONIC SYSTEM 2

_____ **1.** The volumetric flow rate through terminal device 1 is _____ gpm.

_____ **2.** The volumetric flow rate through terminal device 2 is _____ gpm.

_____ **3.** The volumetric flow rate through terminal device 3 is _____ gpm.

_____ **4.** The volumetric flow rate through terminal device 4 is _____ gpm.

_____ **5.** The volumetric flow rate through terminal device 5 is _____ gpm.

6. In a hydronic system, what is the water flow rate based on?

7. When using a pipe sizing chart to size the piping, which size is used if the required pipe size falls between two sizes on the chart?

8. How is the friction head in a piping loop found?

Name _____ Date _____

Calculations

_____ **1.** A terminal device transfers 45,000 Btu/hr. The temperature of the water decreases 6°F as it flows through the terminal device. The required water flow rate is _____ gpm.

_____ **2.** If water enters a terminal device at 189°F and leaves at 72.6°F, what is the temperature difference through the terminal device?

_____ **3.** The pressure drop through a piping system is 3.5′ of head. The water flows through seven terminal devices that have a pressure drop of 1.6′ of head per device. The water also flows through a boiler with a pressure drop of 1.45′ of head. The total system pressure drop is _____′ of head.

_____ **4.** A terminal device transfers 12,000 Btu/hr. The temperature difference of the water as it flows through the terminal device is 10°F. The required water flow rate is _____ gpm.

_____ **5.** The pressure drop through a piping system is 2.5′ of head. The water must flow through twelve terminal devices with a pressure drop of 2.2′ of head per device. The water also flows through a boiler with a pressure drop of 2.7′ of head. The total system pressure drop is _____′ of head.

Short Answer

1. Why must construction details be shown on a layout drawing?

2. Why are design static pressure drops of 300 mpf or 2.5′ of head used for small- to medium-size pipe applications?

3. Define design static pressure drop.

4. Define pressure drop.

5. Why does pipe sizing depend on the material of the pipe used?

Completion

PIPE SIZING CHART*									
Friction Loss§	Nominal Pipe Size#								Friction Loss**
	³⁄₈	¹⁄₂	⁵⁄₈	³⁄₄	1	1¹⁄₄	1¹⁄₂	2	
100	—	.53	.96	1.44	3.1	5.3	8.5	18.2	.83
125	—	.59	1.05	1.63	3.5	6.0	9.6	20.5	1.04
150	—	.65	1.15	1.79	3.8	6.6	10.6	22.6	1.25
175	—	.71	1.26	1.95	4.2	7.3	11.5	24.6	1.46
200	.41	.76	1.35	2.10	4.5	7.8	12.4	26.5	1.67
225	.44	.81	1.44	2.24	4.7	8.3	13.2	28.3	1.88
250	.46	.86	1.53	2.36	5.0	8.7	14.0	30.0	2.08
275	.48	.91	1.61	2.49	5.3	9.2	14.7	31.5	2.29
300	.51	.95	1.68	2.61	5.6	9.6	15.3	33.0	2.50
325	.53	.99	1.75	2.70	5.8	10.0	16.0	34.5	2.71
350	.56	1.03	1.82	2.82	6.1	10.5	16.7	35.8	2.92
375	.58	1.07	1.90	2.93	6.3	10.8	17.3	37.0	3.13
400	.59	1.11	1.96	3.05	6.5	11.1	18.0	38.3	3.33
425	.61	1.15	2.03	3.15	6.7	11.5	18.5	39.6	3.54
450	.63	1.18	2.10	3.24	6.9	11.9	19.1	40.9	3.75

* in gpm
§ in mpf
\# in inches
** in ft/100′

Refer to Pipe Sizing Chart on page 152.

_____ **1.** A piping system is designed with 300 mpf pressure drop and 15 gpm water flow rate. What size pipe should be used?

_____ **2.** A 1″ pipe has a pressure drop of 300 mpf. The water flow rate is _____ gpm.

_____ **3.** A 1¹⁄₂″ pipe has a water flow rate of 15.3 gpm. The pressure drop is _____′ per 100′ of pipe.

_____ **4.** A piping system is designed with 350 mpf pressure drop and 10 gpm water flow rate. What size pipe should be used?

_____ **5.** A ⁵⁄₈″ pipe has a water flow rate of 1.44 gpm. The pressure drop is _____′ per 100′ of pipe.

_____ **6.** A ¹⁄₂″ pipe has a pressure drop of 400 mpf. The water flow rate is _____ gpm.

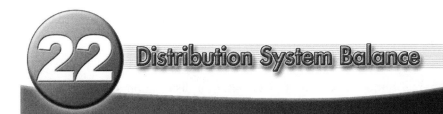

Name _____ Date _____

Completion

_____ **1.** The air flow rate for a blower is a function of the blower wheel diameter, blower wheel _____, and the system pressure drop.

_____ **2.** An anemometer measures air _____ or air flow rate.

_____ **3.** A(n) _____ measures the speed of a shaft or wheel.

_____ **4.** In a(n) _____ system, the air flow rate in each building space remains constant, but the temperature of the supply air varies.

_____ **5.** The air flow rate in ductwork is measured with a(n) _____ or a pitot tube and a manometer.

_____ **6.** A(n) _____ measures the velocity of air flowing out of a register.

_____ **7.** In a one-pipe _____ system, the water flow rate through each terminal device is the total water flow rate for the loop.

_____ **8.** In a(n) _____ system, the air flow rate in the building spaces varies, but the temperature of the supply air remains constant.

_____ **9.** _____ valves are used to adjust the pressure drop for the proper water flow rate in each branch of a piping system.

_____ **10.** Balancing, mixing, and _____ valves are used to balance a hydronic distribution system.

True-False

T   F   **1.** When reading the velocity on an anemometer, the velocity and size of the duct are used to calculate the air flow rate.

T   F   **2.** Distribution systems are usually installed as they are designed.

T   F   **3.** The inside tube of a pitot tube senses static pressure and the outside tube senses total pressure when measuring air flow rate.

T   F   **4.** Air flow rate may be found using the heating capacity and the temperature difference of the air.

T   F   **5.** Blower wheel speed is directly proportional to the diameter of the two sheaves.

T   F   **6.** The blower wheel of a direct drive blower is mounted on the motor shaft.

T   F   **7.** Balancing dampers are used to adjust the air flow rate in a duct.

T   F   **8.** In a two-pipe system, the water flow rate may be different through each terminal device.

T   F   **9.** Balancing valves are sized by pressure drop.

T   F   **10.** The water flow rate through each terminal device in a distribution system is based on the amount of heat required from the terminal device.

T   F   **11.** The water flow rate of a circulating pump in a distribution system is measured when the balancing valves are partially closed.

T   F   **12.** Variable-volume dampers are dampers that control the air at terminal devices in a VAV system.

Short Answer

1. How is the air flow rate of a belt drive blower changed?

2. Why is a forced-air distribution system balanced?

3. Define pressure drop coefficient.

4. How is the air flow rate balanced in a duct?

5. How is the pressure drop in a hydronic distribution system found?

Name _____ Date _____

Activity — Forced-air System Balance

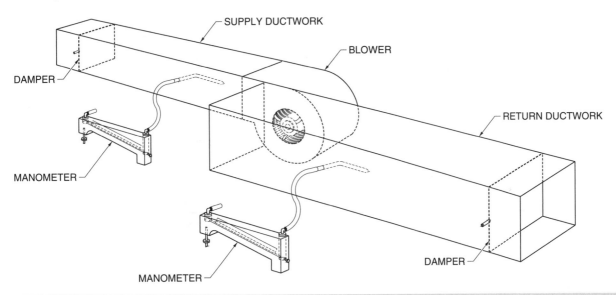

Obtain or construct a duct section with 10' of supply and return ductwork on the blower. The duct section should have balancing dampers near the ends. Manometer taps should be located near the blower inlet and outlet. Turn the blower ON. Measure and record the inlet and outlet static pressure using the manometers with both dampers open, partially closed, and closed. Add the two static pressure values by disregarding the signs. The sum is the total pressure drop in the ductwork against which the blower operates.

	Damper Position		
	Open	**Partially closed**	**Closed**
Inlet static pressure			
Outlet static pressure			
Total pressure drop			

1. Is the pressure increase across a blower the same as the total pressure drop in a system? Explain.

2. How does partially closing a balancing damper simulate adding duct to the system?

3. Can the inlet and outlet static pressures help determine the air flow rate across the blower?

Obtain a blower performance table or graph for the blower. Measure and record the blower wheel speed with a tachometer or calculate the speed using the motor wheel speed and blower and motor sheave sizes. Use the pressure drops and blower wheel speed to determine the air flow rate of the system at the three damper positions from the blower performance table or

	Damper Position		
	Open	Partially closed	Closed
Pressure drop			
Blower speed			
Air flow rate			

4. What is the speed of the blower wheel?

Measure and record the free area of the ductwork with both balancing dampers open, partially closed, and closed. Using a pitot tube, measure and record the velocity pressure in one of the ducts at the same damper position. Calculate the velocity of the air and air flow rate in the duct at each setting.

	Damper Position		
	Open	Partially closed	Closed
P_v			
Duct size			
Air flow rate			

5. Does the measured air flow rate differ from the air flow rate found using a blower performance table or graph? Why or why not?

6. The pitot tube must transverse accurately. Is this important? Was this accomplished?

7. What is the relationship between the static pressure in the duct and velocity pressure?

Place a supply register on the end of the supply ductwork. Make sure the register is large enough to handle all the air delivered by the blower. Use catalogs and specifications sheets to obtain data for the register. Use a velometer to measure and record the velocity of the air out of the register with both balancing dampers open, partially closed, and closed. Use the free area of the register given in the catalog to calculate the air flow rate delivered at each damper setting.

	Damper Position		
	Open	**Partially closed**	**Closed**
Velocity			
Free area			
Air flow rate			

8. What effect does the velocity of the air have on the pattern of the air leaving the damper?

9. In a typical application, should the face velocity from the dampers be the same as the velocity of the air in the supply duct? Explain.

Activity — Hydronic System Balance

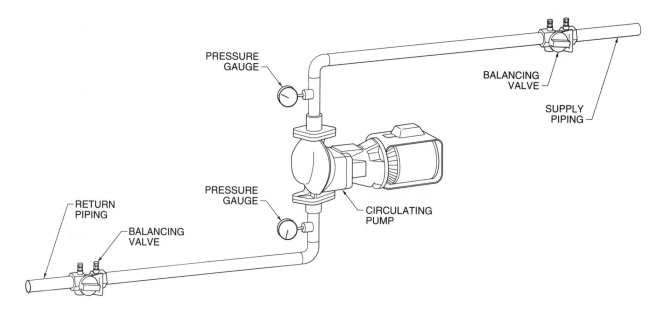

Obtain or construct a piping section with 10′ of supply and return piping on the circulating pump. Install balancing valves with service ports near the ends of the piping section. Install service ports near the circulating pump inlet and outlet. Install pressure gauges in the service ports near the circulating pump. Turn the pump ON. Measure and record the pump inlet and outlet pressures with both balancing valves open, partially closed, and closed. Add the two values by disregarding the signs. The sum is the total pressure drop in the pipe section against which the pump operates.

	Valve Position		
	Open	**Partially closed**	**Closed**
Inlet pressure			
Outlet pressure			
Total pressure			

1. Is the pressure increase across a circulating pump the same as the total pressure drop in a piping system? Explain.

2. How does partially closing a balancing valve simulate adding pipe to the system?

Name _____   Date _____

Calculations

_____ **1.** A duct section is 32″ wide and 16″ high. A velocity pressure reading of .22″ WC is taken with a pitot tube and manometer. The air flow rate is _____ cfm.

_____ **2.** A blower has a 6.5″ blower sheave and a 4.25″ motor sheave. The motor turns at 1725 rpm. If the blower produces an air flow rate of 1250 cfm, what size motor sheave is required to produce an air flow rate of 1375 cfm?

_____ **3.** A blower produces 1100 cfm at a speed of 950 rpm. To produce 1150 cfm, the blower must turn at a speed of _____ rpm.

_____ **4.** The output rating of a boiler is 230,000 Btu/hr. The temperature of the water leaving the boiler is 197.6°F and the temperature of the water entering the boiler is 181.2°F. The water flow rate is _____ gpm.

_____ **5.** An air conditioner has an output rating of 49,500 Btu/hr. The enthalpy of the air entering the air conditioner is 33.2 Btu/lb and the enthalpy of the air leaving the air conditioner is 26.6 Btu/lb. The air flow rate is _____ cfm.

_____ **6.** A furnace has an output rating of 80,000 Btu/hr. The temperature increase of the air through the furnace is 86.7°F. The air flow rate is _____ cfm.

_____ **7.** A velocity pressure reading of .35″ WC is taken with a pitot tube and manometer. The velocity of the air is _____ fpm.

_____ **8.** What is the cross-sectional area of a duct that is 32″ wide and 14″ high?

_____ **9.** A balancing valve has a pressure drop coefficient of 18. The pressure drop across the valve is 3.62 psi. The flow rate through the valve is _____ gpm.

_____ **10.** A duct section is 20″ wide and 8″ high. A velocity pressure reading of .30″ WC is taken with a pitot tube and manometer. The air flow rate is _____ cfm.

_____ **11.** A blower has a 10″ blower sheave and a 6″ motor sheave. The motor turns at 900 rpm. If the blower produces an air flow rate of 1050 cfm, what size motor sheave is required to produce an air flow rate of 1400 cfm?

_____ **12.** The output rating of a boiler is 125,000 Btu/hr. The temperature difference in the system is 18.2°F. The water flow rate is _____ gpm.

_____ **13.** A blower produces 1500 cfm at a speed of 800 rpm. To produce 1800 cfm, the blower must turn at a speed of _____ rpm.

_____ **14.** A refrigeration system has an output rating of 97,000 Btu/hr. The enthalpy of the air entering the air conditioner is 27.8 Btu/lb and the enthalpy of the air leaving the air conditioner is 19.5 Btu/lb. The air flow rate is _____ cfm.

Short Answer

1. Why must an air or water distribution system be balanced?

2. Where is the data concerning the quantity of air required to balance a system found?

3. When should a distribution system be balanced?

4. How is the speed of a direct drive blower changed?

5. How is a distribution system that contains a direct drive circulating pump balanced?

6. Describe a pitot tube.

7. What two methods are used to determine water flow rate in a hydronic piping system?

8. List the three parts of a forced-air distribution system that are adjusted when balancing the system.

Name _____ Date _____

T F **1.** A relative humidity of 50% feels comfortable to most people.

T F **2.** The presence of pollutant sources is an important factor influencing indoor air quality.

T F **3.** The higher the relative humidity, the faster the rate of evaporation.

_____ **4.** _____ is the amount of moisture in the air.

_____ **5.** The air in a forced-air system is cleaned by _____ placed in return air ductwork.

T F **6.** Tagout is the process of removing the source of electrical power and installing a lock that prevents the power from being turned ON.

T F **7.** A lockout/tagout may be removed by anyone who needs to operate the equipment.

_____ **8.** A(n) _____ is a person who has special knowledge, training, and experience in the installation, programming, maintenance, and troubleshooting of HVAC equipment.

_____ **9.** The _____ signal word indicates a potentially hazardous situation which may result in minor or moderate injury.

_____ **10.** _____ is gear worn by HVAC or maintenance technicians to reduce the possibility of injury in the work area.

T F **11.** Heat is transferred by conduction, convection, and radiation.

T F **12.** Heat always flows from a hot object to a cold object.

_____ **13.** _____ heat is heat identified by a change of state of a material.

_____ **14.** Incomplete combustion occurs if there is not enough _____ supplied for combustion.

_____ **15.** Fossil fuels used in boilers include coal, _____, and natural gas.

T F **16.** LP gas is a mixture of propane, butane, and manufactured gas.

T F **17.** Nitrogen passes through a normal combustion process unchanged.

T F **18.** The amount of air required for complete combustion depends on the type of fuel used.

T F **19.** In a parallel circuit, current may be flowing in one part of the circuit even though another part of the circuit is turned OFF.

T F **20.** A transformer is an electric device that changes DC voltage from one level to another.

T F **21.** The total current in a parallel circuit equals the sum of the current through all the loads.

_____   **22.** A(n) _____ circuit has two or more components connected so that there is more than one path for current flow.

_____   **23.** _____ current flow is current flow from positive to negative.
    A. Direct   C. Electron
    B. Alternating   D. Conventional

_____   **24.** A _____ is a switch that interrupts the supply of electric power from motors and machines.
    A. disconnect   C. full-wave rectifier
    B. half-wave rectifier   D. transformer

T   F   **25.** A change in one property of air has no effect on the other properties.

T   F   **26.** A change in the wet bulb temperature of air indicates a change in the moisture content.

T   F   **27.** When moisture is added to air by a humidifier, the dry bulb temperature does not change.

T   F   **28.** The volume of air increases as air cools.

_____   **29.** Enthalpy is the sum of latent heat and _____ heat.

_____   **30.** A sling psychrometer is an instrument used for measuring _____ and consists of wet bulb and dry bulb thermometers mounted on a base.

_____   **31.** A _____ igniter is a device that uses a small piece of silicon carbide that glows when electric current passes through it.
    A. hot surface   C. glow plug
    B. silicon   D. furnace

_____   **32.** Thermostat _____ is the reduction in heating setpoint when occupants are asleep at night or when the space is unoccupied.
    A. control   C. adjustment
    B. setback   D. tuning

T   F   **33.** In a furnace, the products of combustion and heated air are kept completely separate.

T   F   **34.** Registers return air to the system blower.

_____   **35.** A condensate return system is a system used to return condensate back to the _____ after all the useful heat has been removed.

_____   **36.** The _____ of a boiler contains the hot products of combustion and the waterside contains water.

_____   **37.** _____ are components directly attached to a boiler and boiler devices that are required for the operation of the boiler.

_____   **38.** _____ heating is heat transfer that occurs when currents circulate between warm and cool regions of a fluid.

T   F   **39.** The pressure rating of a safety valve is the amount of steam that the safety valve is capable of venting at rated pressure.

T   F   **40.** A gauge glass is a tubular glass column that indicates the water level in the boiler.

_____   **41.** Low-pressure chillers use _____ compressors.

_____   **42.** A(n) _____ is a component in a chiller that transfers the heat from the refrigerant to a cooling medium, normally water but sometimes air.

_____ **43.** A(n) _____ is the component in the chiller system that transfers heat from the water to the liquid refrigerant.

T F **44.** An absorption refrigeration system produces a refrigeration effect with mechanical equipment.

T F **45.** If the pressure on a liquid increases, the boiling point temperature decreases.

T F **46.** A gas cylinder must have a protective cap over the valve while the cylinder is being moved.

T F **47.** Fluorocarbon-based refrigerants can break down in the presence of an open flame and generate poisonous phosgene gas.

T F **48.** Lockout/tagout regulations are optional when a trained technician is working on refrigeration and air conditioning equipment.

_____ **49.** Technicians must not vent, release, or dispose of any substance used as a refrigerant in a manner which permits such a substance to enter the _____.

_____ **50.** Refrigerant _____ is when refrigerant is processed to new product specifications by means that may include distillation.

    A. removal   C. reclaiming
    B. recycling   D. recovery

T F **51.** High compression ratio indicates an efficiently operating refrigeration system.

T F **52.** The total heat rejected by a refrigeration system must equal the heat absorbed in the evaporator plus the heat of compression.

T F **53.** Superheat is sensible heat added to a substance after it has changed state.

_____ **54.** _____ is the total heat contained in a substance.

_____ **55.** Refrigeration effect is the amount of heat absorbed by a refrigerant in the _____ of a refrigeration system.

    A. evaporator   C. compressor
    B. expansion device   D. condenser

_____ **56.** The pressure drop created by an expansion device is accompanied by a(n) _____ drop that enables the system to produce a cooling effect.

T F **57.** A capillary tube expansion device can handle significant changes in the cooling load of a refrigeration system.

T F **58.** The thermostatic expansion valve and the automatic expansion valve control the temperature of the refrigerant in a system.

T F **59.** The expansion device and compressor suction pressure maintain the low pressure in the low-pressure side of a mechanical compression refrigeration system.

T F **60.** A float valve controls liquid level.

T F **61.** As a refrigerant flows through a condenser coil, heat is transferred from the air moving across the coil to the refrigerant flowing through the coil.

T F **62.** During the compression stroke of a reciprocating compressor, the pressure in the cylinder increases and closes the suction valve.

T F **63.** To achieve higher pressures with centrifugal compressors, multistage units are used.

_____ **64.** Two kinds of rotary compressors used in air conditioning systems are stationary vane and _____ vane compressors.

_____ **65.** A(n) _____ compressor is a compressor that has all of the components and the motor sealed in a metal housing.

T   F   **66.** A four-pipe hydronic distribution system allows hot or cold water to flow to any terminal device.

T   F   **67.** A circulating pump on a hydronic air conditioning system removes condensate from the system.

T   F   **68.** A combination unit is an HVAC unit that contains the components for cooling and heating in one sheet metal cabinet.

_____ **69.** The two most common evaporating mediums used in air conditioning systems are water and _____.

_____ **70.** Air conditioners use a mechanical compression or a(n) _____ process.

T   F   **71.** As a heat pump switches from the cooling mode to the heating mode, the pressure in the two sides of the system remains the same.

T   F   **72.** A capillary tube can be used as an expansion device on a heat pump because it produces a pressure decrease in the refrigerant regardless of the direction of flow.

T   F   **73.** A thermostatic expansion valve allows refrigerant flow in two directions.

_____ **74.** The direction of refrigerant flow in a heat pump is controlled by a(n) _____ valve.

_____ **75.** A(n) _____ valve is a valve that pierces a refrigerant line.

T   F   **76.** Temperature limit switches are used to start the burner on a heating unit.

T   F   **77.** A field interface device is an electronic device that sends commands to the CPU of a central-direct digital control system.

T   F   **78.** A flow switch is a device that surrounds a wire and detects the electromagnetic field caused by electricity passing through the wire.

_____ **79.** Relays use a _____ to create magnetism to open or close one or more sets of contacts.

A. coil   C. fuse
B. thermostat   D. transformer

_____ **80.** A stack switch senses _____.

A. carbon dioxide levels   C. flue gas temperature
B. carbon monoxide levels   D. presence of smoke

T   F   **81.** Infiltration occurs when air flows through a furnace filter.

_____ **82.** _____ is the process that occurs when outdoor air is brought into a building.

T   F   **83.** Any device that produces heat inside a building must be compensated for by a cooling load.

T   F   **84.** Total heat loss or gain for a building is the sum of all the room or zone subtotals.

_____ **85.** For residential cooling applications, heat gain from people is _____ Btu/hr of sensible heat per person with a latent heat allowance added.

_____ **86.** A summer indoor design temperature of _____ °F is usually selected to provide comfort for a normally clothed person during summer.

_____ **87.** _____ furnaces are used where the supply ductwork is located above the furnace.

_____ **88.** _____ rating is the amount of heat a furnace produces in 1 hour.

T   F   **89.** The cooling capacity of an air conditioner must be as great as or greater than the calculated cooling load of the building.

T   F   **90.** Absorption chillers are used in buildings of all sizes because the chillers are compact and can be adapted to all conditions.

T   F   **91.** Static pressure drop is the decrease in air pressure caused by friction between the air moving through a duct and the internal surfaces of a duct.

T   F   **92.** To function most efficiently, the aspect ratio of a rectangular duct should not be more than 3 to 4 times the height.

T   F   **93.** Water pressure can be expressed in feet of head and pounds per square inch.

T   F   **94.** Volumetric flow rate of water is measured in feet per minute.

T   F   **95.** The pipe size remains the same throughout a one-pipe series system because the flow rate remains the same.

_____ **96.** An anemometer measures air _____ or air flow rate.

_____ **97.** A(n) _____ measures the speed of a shaft or wheel.

_____ **98.** A(n) _____ measures the velocity of air flowing out of a register.

T   F   **99.** The blower wheel of a direct drive blower is mounted on the motor shaft.

T   F **100.** Balancing dampers are used to adjust the air flow rate in a duct.

Name _____   Date _____

_____ **1.** _____°C equals 80°F.

_____ **2.** If 26 kW of electrical energy is converted to heat, how much heat is produced?

_____ **3.** A boiler with an input rating of 240,000 Btu/hr has an efficiency rating of 85%. The output rating of the boiler is _____ Btu/hr.

_____ **4.** The specific heat of water is 1. How much heat is required to raise the temperature of 800 lb of water 20°F?

_____ **5.** The blower in a furnace has a 4″ sheave. The motor that turns the blower has a 2″ sheave. If the blower wheel turns at 1200 rpm, the speed of the blower motor is _____ rpm.

_____ **6.** What is the absolute pressure for a gauge pressure of 15 psig?

_____ **7.** How much heat is produced when 373 gal. of Grade No. 2 fuel oil is burned at 100% efficiency? Use the standard heating value for Grade No. 2.

_____ **8.** A refrigeration system uses 157,000 Btu/hr of energy to produce 123,000 Btu/hr of cooling. The COP of the system is _____.

_____ **9.** The enthalpy of the refrigerant that leaves the compressor in a mechanical compression refrigeration system is 976 Btu/lb. The enthalpy of the refrigerant that enters the compressor is 923 Btu/lb. The heat of compression is _____ Btu/lb.

_____ **10.** Gypsum board ½″ thick has a C factor of 2.22. If the temperature difference through the board is 36°F, the amount of heat conducted through 780 sq ft of gypsum board is _____ Btu/hr.

_____ **11.** A refrigeration system has a high-pressure side of 190 psig and a low-pressure side of 75 psig. The compression ratio is _____.

_____ **12.** A gas-fired furnace burns natural gas at a rate of 80 cu ft/hr. The input rating of the furnace is _____ Btu/hr. Use the standard heating value for natural gas.

_____ **13.** What is the current in a circuit with a 141 Ω solenoid and 24 V power supply?

_____ **14.** An air-cooled evaporator produces a refrigeration effect of 195 Btu/lb. The mass flow rate of the refrigerant is 14.9 lb/min. The cooling capacity of the evaporator is _____ Btu/hr.

_____ **15.** How much heat is produced when 39 cu ft of natural gas is burned at 100% efficiency? Use the standard heating value for natural gas.

_____ **16.** The pistons in a compressor have diameters of 2″ and stroke lengths of 2.5″. The volumetric capacity of one cylinder is _____ cu in.

_____ **17.** An electric motor turns at 1725 rpm, has a 3.5″ motor sheave, and is connected to a blower wheel with a 6.5″ blower sheave. Find the blower wheel speed.

_____ 18. A refrigerant enters the evaporator of a refrigeration system with an enthalpy of 67 Btu/lb and leaves with an enthalpy of 839 Btu/lb. The refrigeration effect of the evaporator is _____ Btu/lb.

_____ 19. The cooling capacity of an air-cooled evaporator is 240,000 Btu/hr. The actual refrigeration effect of the system is _____ Btu/hr.

_____ 20. An evaporator produces a refrigeration effect of 80.6 Btu/lb. The mass flow rate of the refrigerant that provides a 77,000 Btu/hr cooling capacity is _____ lb/min.

_____ 21. A refrigeration system has a refrigerant mass flow rate of 25.3 lb/min. The specific volume of the refrigerant is .655 cu ft/lb. The volumetric flow rate of the refrigerant is _____ cfm.

_____ 22. A condenser rejects 52.8 Btu/lb of refrigerant. Find the heat rejection rate of the condenser if the mass flow rate of the refrigerant is 19.3 lb/min.

_____ 23. A furnace with an input rating of 99,000 Btu/hr has an efficiency rating of 92%. The output rating of the furnace is _____ Btu/hr.

_____ 24. _____°C equals 105°F.

_____ 25. A compressor has two cylinders. The pistons in the cylinders have diameters of 2″ and stroke lengths of 1.5″. What is the volumetric capacity of the compressor if the compressor turns at 1750 rpm?

_____ 26. A condenser coil has a heat rejection rate of 73,000 Btu/hr. The condenser blower moves 2000 cfm of air across the coil. The temperature difference of the air leaving the condenser coil is _____°F.

_____ 27. A 17.5 ton air conditioner produces _____ Btu/hr of cooling.

_____ 28. A refrigerant enters the condenser of a refrigeration system with an enthalpy of 1163 Btu/lb and leaves with an enthalpy of 77 Btu/lb. The heat of rejection is _____ Btu/lb.

_____ 29. If .8 kW of electrical energy is converted to heat, how much heat is produced?

_____ 30. The wall of a building is 40′ wide and 9′ high. The gross wall area is _____ sq ft.

_____ 31. What is the total resistance in a circuit containing a 160 Ω solenoid, an 82 Ω fan relay, and a 28 Ω crankcase heater wired in parallel?

_____ 32. How much heat is produced when 9 lb of coal is burned at 100% efficiency? Use the standard heating value for coal.

_____ 33. If a building has 27 occupants and 30 cfm of air per person is required for ventilation, what is the volumetric flow rate of ventilation air?

_____ 34. Cement plaster ¾″ thick has a C factor of 6.66. If the temperature difference through the plaster is 28°F, the amount of heat conducted through 320 sq ft of plaster is _____ Btu/hr.

_____ 35. The blower in a duct turns at 1450 rpm. The blower motor turns at 3750 rpm. Find the diameter of the motor sheave if the blower sheave is 6″.

_____ 36. A refrigeration system has a high-pressure side of 187 psig and a low-pressure side of 60 psig. The compression ratio is _____.

_____ 37. If the gross wall area of a building is 270 sq ft and the area of window and door openings is 62 sq ft, what is the net wall area?

_____ **38.** A water-cooled condenser has a heat rejection rate of 104,000 Btu/hr. The volumetric flow rate of water through the condenser is 14 gpm. The temperature increase of the water leaving the condenser is _____°F.

_____ **39.** An evaporator produces a refrigeration effect of 127.2 Btu/lb. The mass flow rate of the refrigerant that provides a 167,000 Btu/hr cooling capacity is _____ lb/min.

_____ **40.** The specific heat of air is .24. How much heat is required to raise the temperature of 30 lb of air 50°F?

_____ **41.** A condenser rejects 113.5 Btu/lb of refrigerant. Find the heat rejection rate of the condenser if the mass flow rate of the refrigerant is 21.9 lb/min.

_____ **42.** A 90 ton chiller produces _____ Btu/hr of cooling.

_____ **43.** If a winter indoor design temperature of 70°F is used for an application and the outdoor design temperature is 23°F, the design temperature difference is _____°F.

_____ **44.** The cooling load on a building is 27,629 Btu/hr. A 36,000 Btu/hr air conditioner is chosen. The cooling equipment ratio is _____.

_____ **45.** A refrigeration system uses 80,000 Btu/hr of energy to produce 60,000 Btu/hr of cooling. What is the COP of the system?

_____ **46.** A room has an area of 480 sq ft. The blower in a furnace circulates 700 cfm of air. Does the air flow rate create a draft?

_____ **47.** A compressor has four cylinders. The pistons in the cylinders have diameters of 2.25″ and stroke lengths of 2″. If the compressor turns at 1250 rpm, the volumetric capacity of the compressor is _____ cfm.

_____ **48.** _____°F equals 15°C.

_____ **49.** What is the voltage drop created by a 22 Ω resistor in a series circuit with a 46 Ω relay coil and a 62 Ω temperature sensor? The circuit has a 24 V power supply.

_____ **50.** The cooling capacity of an air-cooled evaporator is 144,000 Btu/hr. The actual refrigeration effect of the system is _____ Btu/hr.

_____ **51.** How much heat is produced when 16 cu ft of propane is burned at 100% efficiency? Use the standard heating value for propane.

_____ **52.** The velocity of the air entering a fitting is 2000 fpm. The velocity pressure in the fitting is _____″ WC.

_____ **53.** A grill in a forced-air distribution system must handle 600 cfm of air at a face velocity of 400 fpm. The required free area is _____ sq in.

_____ **54.** What is the absolute pressure for a gauge pressure of 49 psig?

_____ **55.** A duct has a design static pressure drop of .1″ WC. The duct section is 75′ long. The static pressure drop in the duct section is _____″ WC.

_____ **56.** The floor in a room is 24′ long and 12′ wide. The surface area of the floor is _____ sq ft.

_____ **57.** The pressure drop through a piping system is 2.5′ of head. The water flows through four terminal devices with a pressure drop of 4′ of head per device. The water also flows through a boiler with a pressure drop of 3.2′ of head. The total system pressure drop is _____′ of head.

_____ 58. A building has a total heat loss of 78,945 Btu/hr. The ductwork in the building is located in an unheated crawlspace. The duct loss multiplier is 1.10. The adjusted heat loss due to ductwork is _____ Btu/hr.

_____ 59. The pistons in a compressor have diameters of 1.5″ and stroke lengths of 2″. The volumetric capacity of one cylinder is _____ cu in.

_____ 60. A fuel-oil fired furnace burns Grade No. 2 fuel oil at a rate of 11 gph. The input rating of the furnace is _____ Btu/hr. Use the standard heating value for Grade No. 2.

_____ 61. If a winter indoor design temperature of 70°F is used for an application and the outdoor design temperature is 5°F, what is the design temperature difference?

_____ 62. A balancing valve has a pressure drop coefficient of 12. The pressure drop across the valve is 1.7 psi. The flow rate through the valve is _____ gpm.

_____ 63. A refrigeration system has a water-cooled condenser with a water flow rate of 14 gpm. The temperature difference of the water flowing through the condenser is 19°F. The heat rejection rate of the condenser is _____ Btu/hr.

_____ 64. A refrigerant enters the evaporator of a refrigeration system with an enthalpy of 96 Btu/lb and leaves with an enthalpy of 1142 Btu/lb. The refrigeration effect of the evaporator is _____ Btu/lb.

_____ 65. How many therms of natural gas is required to produce 260,000 Btu?

_____ 66. The pressure drop through a piping system is 2.5′ of head. The water flows through six terminal devices with a pressure drop of 2.3′ of head per device. The water also flows through a boiler with a pressure drop of 5′ of head. The total system pressure drop is _____′ of head.

_____ 67. A grill in a forced-air distribution system must handle 1000 cfm of air at a face velocity of 550 fpm. The required free area is _____ sq in.

_____ 68. What is the voltage drop created by a 36 Ω resistor in series with a 48 Ω temperature sensor in one leg of a parallel circuit that has a 24 V power supply?

_____ 69. A refrigeration system has a mass flow rate of 19 lb/min. The specific volume of the refrigerant is .248 cu ft/lb. What is the volumetric flow rate of the refrigerant?

_____ 70. The refrigeration effect of an air-cooled evaporator is 99 Btu/lb. The mass flow rate of the refrigerant is 16 lb/min. What is the cooling capacity of the evaporator?

_____ 71. The cooling load on a building is 31,720 Btu/hr. A 36,000 Btu/hr air conditioner is chosen. The cooling equipment ratio is _____.

_____ 72. A .5 ton air conditioner produces _____ Btu/hr of cooling.

_____ 73. A room has an area of 1250 sq ft. The blower in a furnace circulates 800 cfm of air. Does the air flow rate create a draft?

_____ 74. A duct has a design static pressure drop of .08″ WC. The duct section is 84′ long. The static pressure drop of the duct section is _____″ WC.

_____ 75. A balancing valve has a pressure drop coefficient of 17. The pressure drop across the valve is 2.1 psi. The flow rate through the valve is _____ gpm.

Appendix

REFRIGERANTS

LOAD CALCULATIONS

TROUBLESHOOTING CHARTS

R-12—PROPERTIES OF SATURATED LIQUID AND SATURATED VAPOR

Temp. °F	Pressure		Volume§	Density#	Enthalpy**		Entropy§§	
	psia	psig	Vapor	Liquid	Liquid	Vapor	Liquid	Vapor
−140	0.25567	29.401*	110.72	104.03	−20.398	62.190	−.055433	.20292
−130	0.41131	29.084*	70.904	103.12	−18.380	63.258	−.049218	.19842
−120	0.64047	28.617*	46.858	102.21	−16.362	64.338	−.043187	.19440
−110	0.96829	27.950*	31.857	101.29	−14.341	65.430	−.037326	.19081
−100	1.4252	27.019*	22.220	100.36	−12.317	66.530	−.031620	.18760
− 90	2.0473	25.753*	15.861	99.429	−10.287	67.638	−.026058	.18474
− 85	2.4332	24.967*	13.509	98.958	−9.2699	68.194	−.023326	.18343
− 80	2.8765	24.065*	11.563	98.486	−8.2506	68.751	−.020626	.18219
− 75	3.3834	23.033*	9.9442	98.012	−7.2292	69.310	−.017956	.18102
− 70	3.9604	21.858*	8.5912	97.535	−6.2054	69.869	−.015314	.17991
− 65	4.6144	20.526*	7.4545	97.056	−5.1789	70.429	−.012700	.17887
− 60	5.3526	19.023*	6.4950	96.575	−4.1497	70.990	−.010112	.17789
− 55	6.1826	17.333*	5.6813	96.091	−3.1174	71.550	−.007549	.17697
− 50	7.1124	15.440*	4.9884	95.605	−2.0818	72.111	−.005010	.17609
− 45	8.1502	13.327*	4.3957	95.116	−1.0427	72.672	−.002494	.17527
− 40	9.3045	10.977*	3.8868	94.624	0.0	73.232	.0	.17450
− 38	9.8008	9.9666*	3.7035	94.426	0.4182	73.456	.000992	.17420
− 36	10.318	8.9139*	3.5307	94.228	0.8370	73.679	.001980	.17391
− 34	10.856	7.8180*	3.3677	94.029	1.2564	73.903	.002965	.17363
− 32	11.416	6.6777*	3.2138	93.830	1.6766	74.127	.003948	.17335
− 30	11.999	5.4916*	3.0684	93.631	2.0974	74.350	.004927	.17309
− 28	12.604	4.2586*	2.9310	93.431	2.5189	74.574	.005903	.17282
− 26	13.234	2.9773*	2.8011	93.230	2.9411	74.797	.006876	.17257
− 24	13.887	1.6466*	2.6782	93.029	3.3641	75.020	.007845	.17232
− 22	14.566	0.2651*	2.5618	92.827	3.7877	75.242	.008813	.17207
− 21.62	14.696	0.0	2.5407	92.789	3.8673	75.284	.008994	.17203
− 20	15.270	0.5739	2.4516	92.625	4.2121	75.465	.009777	.17184
− 18	16.000	1.3043	2.3471	92.422	4.6372	75.687	.010738	.17160
− 16	16.758	2.0615	2.2481	92.219	5.0631	75.909	.011697	.17138
− 14	17.542	2.8463	2.1541	92.015	5.4898	76.131	.012653	.17116
− 12	18.355	3.6592	2.0650	91.810	5.9173	76.353	.013606	.17094
− 10	19.197	4.5011	1.9803	91.605	6.3456	76.574	.014557	.17073
− 8	20.069	5.3725	1.8999	91.399	6.7747	76.795	.015505	.17053
− 6	20.970	6.2742	1.8234	91.192	7.2046	77.015	.016451	.17033
− 4	21.903	7.2067	1.7507	90.985	7.6354	77.236	.017394	.17014
− 2	22.867	8.1710	1.6815	90.777	8.0670	77.456	.018335	.16995

* in in. Hg. vacuum
§ in cu ft/lb
# in lb/cu ft
** in Btu/lb
§§ in Btu/lb × °R

continued

continued

R-12—PROPERTIES OF SATURATED LIQUID AND SATURATED VAPOR

Temp. °F	Pressure		Volume§	Density#	Enthalpy**		Entropy§§	
	psia	psig	Vapor	Liquid	Liquid	Vapor	Liquid	Vapor
0	23.863	9.1675	1.6157	90.569	8.4995	77.675	.019274	.16976
2	24.893	10.197	1.5530	90.360	8.9329	77.895	.020210	.16958
4	25.956	11.260	1.4933	90.150	9.3671	78.114	.021144	.16941
6	27.054	12.358	1.4364	89.939	9.8023	78.332	.022075	.16924
8	28.187	13.491	1.3821	89.728	10.238	78.550	.023005	.16907
10	29.356	14.660	1.3303	89.516	10.675	78.768	.023932	.16891
12	30.561	15.865	1.2809	89.303	11.113	78.985	.024857	.16875
14	31.804	17.108	1.2337	89.089	11.552	79.202	.025780	.16860
16	33.085	18.389	1.1887	88.875	11.992	79.418	.026701	.16845
18	34.405	19.709	1.1456	88.659	12.433	79.633	.027620	.16830
20	35.765	21.069	1.1045	88.443	12.874	79.849	.028537	.16816
25	39.341	24.645	1.0093	87.899	13.983	80.384	.030821	.16782
30	43.182	28.486	.92401	87.349	15.098	80.916	.033093	.16751
35	47.300	32.604	.84738	86.793	16.220	81.444	.035354	.16721
40	51.705	37.009	.77838	86.231	17.348	81.968	.037604	.16693
50	61.432	46.736	.65984	85.087	19.625	83.000	.042075	.16642
60	72.462	57.766	.56254	83.912	21.931	84.012	.046508	.16597
70	84.900	70.204	.48204	82.704	24.266	84.998	.050906	.16557
80	98.850	84.154	.41495	81.458	26.633	85.955	.055273	.16520
90	114.42	99.725	.35864	80.168	29.032	86.880	.059613	.16485
100	131.72	117.03	.31106	78.830	31.466	87.766	.063928	.16452
110	150.87	136.18	.27060	77.436	33.937	88.609	.068224	.16420
120	171.99	157.29	.23597	75.978	36.448	89.400	.072504	.16385
130	195.20	180.50	.20615	74.446	39.003	90.133	.076777	.16349
140	220.63	205.93	.18032	72.827	41.608	90.795	.081050	.16307
150	248.43	233.73	.15780	71.106	44.271	91.375	.085335	.16260
160	278.74	264.05	.13804	69.264	47.002	91.856	.089649	.16203
170	311.74	297.04	.12057	67.275	49.818	92.214	.094013	.16134
180	347.61	332.91	.10499	65.103	52.741	92.418	.098461	.16049
190	386.56	371.86	.090963	62.697	55.808	92.419	.10304	.15940
200	428.84	414.14	.078132	59.979	59.075	92.139	.10784	.15796
210	474.75	460.06	.066108	56.806	62.647	91.430	.11299	.15597
220	524.70	510.01	.054263	52.851	66.748	89.942	.11882	.15295
230	579.33	564.63	.040319	46.607	72.294	86.110	.12662	.14665
##233.2	598.3	583.6	.02871	34.83	79.40	79.40	.1368	.1368

* in in. Hg. vacuum
§ in cu ft/lb
# in lb/cu ft

** in Btu/lb
§§ in Btu/lb × °R
Critical point

ASHRAE Handbook — Fundamentals

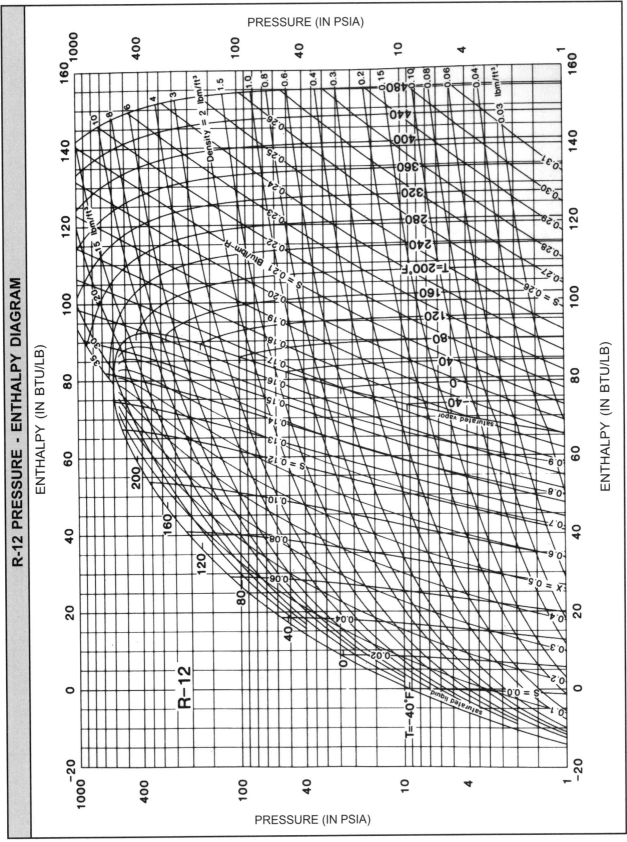

R-12 PRESSURE - ENTHALPY DIAGRAM

PRESSURE (IN PSIA)

ENTHALPY (IN BTU/LB)

R-12

R-22—PROPERTIES OF SATURATED LIQUID AND SATURATED VAPOR

Temp. °F	Pressure		Volume§	Density#	Enthalpy**		Entropy§§	
	psia	psig	Vapor	Liquid	Liquid	Vapor	Liquid	Vapor
−130	0.68858	28.519*	59.170	96.313	−24.388	89.888	−.065456	.28118
−120	1.0725	27.738*	39.078	95.416	−21.538	91.049	−.056942	.27452
−110	1.6199	26.623*	26.578	94.509	−18.738	92.211	−.048818	.26848
−100	2.3802	25.075*	18.558	93.590	−15.980	93.371	−.041046	.26298
− 90	3.4111	22.976*	13.268	92.660	−13.259	94.527	−.033590	.25798
− 80	4.7793	20.191*	9.6902	91.717	−10.570	95.676	−.026418	.25342
− 75	5.6131	18.493*	8.3419	91.241	−9.2346	96.247	−.022929	.25128
− 70	6.5603	16.564*	7.2139	90.761	−7.9050	96.815	−.019500	.24924
− 65	7.6317	14.383*	6.2655	90.278	−6.5802	97.380	−.016128	.24728
− 60	8.8386	11.926*	5.4641	89.791	−5.2593	97.942	−.012808	.24541
− 50	11.707	6.0851*	4.2039	88.807	−2.6263	99.055	−.006316	.24189
− 48	12.361	4.7548*	3.9962	88.608	−2.1007	99.275	−.005039	.24122
− 46	13.042	3.3666*	3.8007	88.408	−1.5753	99.495	−.003770	.24056
− 44	13.754	1.9186*	3.6168	88.208	−1.0501	99.714	−.002507	.23991
− 42	14.495	0.4090*	3.4437	88.007	−0.5250	99.932	−.001250	.23927
− 41.47	14.696	0.0	3.3997	87.954	−0.3865	99.990	−.000920	.23910
− 40	15.268	0.5717	3.2805	87.806	0.0	100.15	.0	.23864
− 38	16.072	1.3763	3.1267	87.604	0.5250	100.37	.001244	.23802
− 36	16.910	2.2138	2.9816	87.401	1.0500	100.58	.002482	.23741
− 34	17.781	3.0852	2.8446	87.197	1.5751	100.80	.003714	.23681
− 32	18.687	3.9914	2.7152	86.993	2.1003	101.01	.004940	.23622
− 30	19.629	4.9333	2.5930	86.788	2.6257	101.22	.006161	.23564
− 28	20.608	5.9119	2.4774	86.582	3.1512	101.44	.007377	.23506
− 26	21.624	6.9283	2.3680	86.375	3.6771	101.65	.008587	.23450
− 24	22.679	7.9832	2.2645	86.168	4.2032	101.86	.009792	.23394
− 22	23.774	9.0778	2.1664	85.960	4.7297	102.07	.010993	.23340
− 20	24.909	10.213	2.0735	85.751	5.2566	102.28	.012188	.23285
− 18	26.086	11.390	1.9854	85.542	5.7840	102.48	.013379	.23232
− 16	27.306	12.610	1.9018	85.331	6.3119	102.69	.014566	.23180
− 14	28.569	13.873	1.8225	85.120	6.8403	102.90	.015748	.23128
− 12	29.877	15.181	1.7472	84.908	7.3693	103.10	.016926	.23077
− 10	31.231	16.535	1.6757	84.695	7.8989	103.30	.018100	.23027
− 8	32.632	17.936	1.6077	84.481	8.4292	103.51	.019270	.22977
− 6	34.081	19.385	1.5430	84.266	8.9603	103.71	.020436	.22928
− 4	35.579	20.883	1.4815	84.051	9.4921	103.91	.021598	.22880
− 2	37.127	22.431	1.4230	83.834	10.025	104.10	.022757	.22832

* in in. Hg vacuum
§ in cu ft/lb
# in lb/cu ft
** in Btu/lb
§§ in Btu/lb × °R

continued

continued

R-22—PROPERTIES OF SATURATED LIQUID AND SATURATED VAPOR

Temp. °F	Pressure		Volume§	Density#	Enthalpy**		Entropy§§	
	psia	psig	Vapor	Liquid	Liquid	Vapor	Liquid	Vapor
0	38.726	24.030	1.3672	83.617	10.558	104.30	.023912	.22785
2	40.378	25.682	1.3141	83.399	11.093	104.50	.025064	.22738
4	42.083	27.387	1.2635	83.179	11.628	104.69	.026213	.22693
6	43.843	29.147	1.2152	82.959	12.164	104.89	.027359	.22647
8	45.658	30.962	1.1692	82.738	12.702	105.08	.028502	.22602
10	47.530	32.834	1.1253	82.516	13.240	105.27	.029642	.22558
12	49.461	34.765	1.0833	82.292	13.779	105.46	.030779	.22515
14	51.450	36.754	1.0433	82.068	14.320	105.64	.031913	.22471
16	53.501	38.805	1.0050	81.843	14.862	105.83	.033045	.22429
18	55.612	40.916	.96841	81.616	15.405	106.02	.034175	.22387
20	57.786	43.090	.93343	81.389	15.950	106.20	.035302	.22345
25	63.505	48.809	.85246	80.815	17.317	106.65	.038110	.22243
30	69.641	54.945	.77984	80.234	18.693	107.09	.040905	.22143
35	76.215	61.519	.71454	79.645	20.078	107.52	.043689	.22046
40	83.246	68.550	.65571	79.049	21.474	107.94	.046464	.21951
45	90.754	76.058	.60258	78.443	22.880	108.35	.049229	.21858
50	98.758	84.062	.55451	77.829	24.298	108.74	.051987	.21767
55	107.28	92.583	.51093	77.206	25.728	109.12	.054739	.21677
60	116.34	101.64	.47134	76.572	27.170	109.49	.057486	.21589
65	125.95	111.26	.43531	75.928	28.626	109.84	.060228	.21502
70	136.15	121.45	.40245	75.273	30.095	110.18	.062968	.21416
80	158.36	143.66	.34497	73.926	33.077	110.80	.068441	.21246
90	183.14	168.44	.29668	72.525	36.121	111.35	.073911	.21077
100	210.67	195.97	.25582	71.061	39.233	111.81	.079400	.20907
110	241.13	226.44	.22102	69.524	42.422	112.17	.084906	.20734
120	274.73	260.03	.19118	67.901	45.694	112.42	.090444	.20554
130	311.66	296.96	.16542	66.174	49.064	112.52	.096033	.20365
140	352.14	337.45	.14300	64.319	52.550	112.47	.10170	.20161
150	396.42	381.72	.12334	62.301	56.177	112.20	.10749	.19938
160	444.75	430.06	.10590	60.068	59.989	111.67	.11345	.19684
170	497.46	482.76	.090228	57.532	64.055	110.76	.11970	.19386
180	554.89	540.19	.075819	54.533	68.504	109.30	.12640	.19018
190	617.52	602.82	.061991	50.703	73.617	106.88	.13399	.18518
200	686.02	671.32	.046923	44.671	80.406	101.99	.14394	.17666
##205.07	723.4	708.7	.03123	32.03	91.58	91.58	.1605	.1605

* in in. Hg. vacuum
§ in cu ft/lb
# in lb/cu ft

** in Btu/lb
§§ in Btu/lb × °R
Critical point

ASHRAE Handbook — Fundamentals

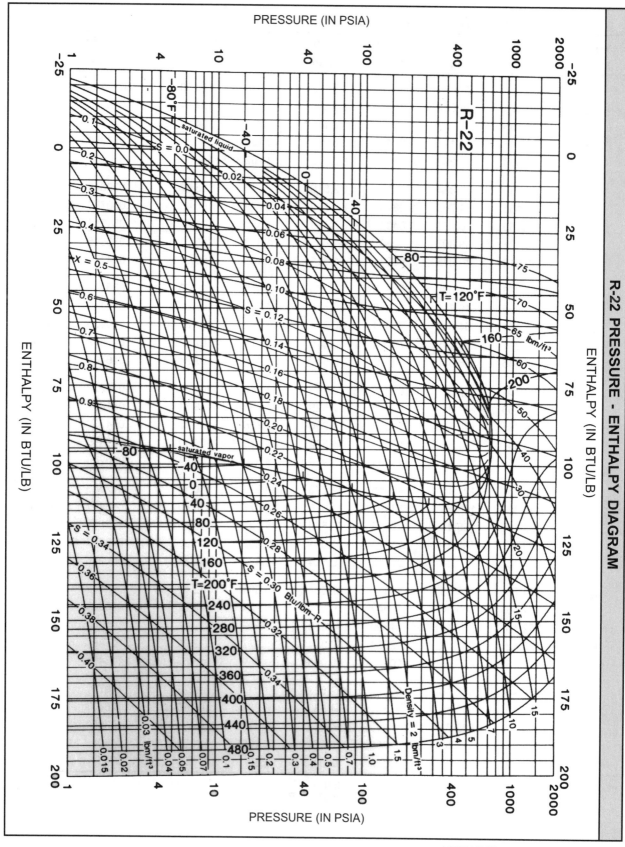

PRESSURE (IN PSIA)

R-22 PRESSURE - ENTHALPY DIAGRAM

ENTHALPY (IN BTU/LB)

ENTHALPY (IN BTU/LB)

PRESSURE (IN PSIA)

R-502—PROPERTIES OF SATURATED LIQUID AND SATURATED VAPOR

Temp. °F	Pressure psia	Pressure psig	Volume§ Vapor	Density# Liquid	Enthalpy** Liquid	Enthalpy** Vapor	Entropy§§ Liquid	Entropy§§ Vapor
−75	7.2807	15.098*	4.9585	95.235	−7.6022	68.964	−.018844	.18020
−70	8.4343	12.749*	4.3241	94.699	−6.5643	69.570	−.016169	.17921
−65	9.7307	10.109*	3.7849	94.159	−5.5105	70.174	−.013488	.17828
−60	11.182	7.1539*	3.3248	93.615	−4.4406	70.777	−.010802	.17740
−55	12.802	3.8567*	2.9306	93.067	−3.3546	71.377	−.008110	.17656
−50	14.602	0.1906*	2.5915	92.514	−2.2525	71.975	−.005412	.17578
−49.75	14.696	0.0	2.5761	92.487	−2.1978	72.004	−.005279	.17574
−48	15.376	0.6804	2.4693	92.292	−1.8071	72.213	−.004331	.17547
−46	16.183	1.4865	2.3541	92.068	−1.3592	72.451	−.003249	.17518
−44	17.022	2.3256	2.2452	91.844	−0.9087	72.689	−.002167	.17489
−42	17.895	3.1986	2.1425	91.620	−0.4556	72.925	−.001084	.17461
−40	18.802	4.1064	2.0453	91.394	0.0	73.162	.0	.17433
−38	19.746	5.0500	1.9535	91.167	0.4582	73.398	.001085	.17406
−36	20.726	6.0303	1.8666	90.940	0.9189	73.633	.002170	.17380
−34	21.744	7.0482	1.7844	90.712	1.3822	73.867	.003256	.17354
−32	22.801	8.1048	1.7066	90.483	1.8481	74.101	.004343	.17329
−30	23.897	9.2010	1.6328	90.253	2.3165	74.335	.005430	.17304
−28	25.034	10.338	1.5629	90.022	2.7874	74.567	.006518	.17280
−26	26.212	11.516	1.4966	89.790	3.2608	74.799	.007607	.17257
−24	27.433	12.737	1.4336	89.557	3.7367	75.031	.008696	.17234
−22	28.697	14.001	1.3739	89.323	4.2152	75.261	.009786	.17211
−20	30.006	15.310	1.3172	89.088	4.6961	75.491	.010876	.17189
−18	31.361	16.665	1.2633	88.853	5.1795	75.720	.011966	.17168
−16	32.762	18.066	1.2120	88.616	5.6654	75.948	.013057	.17147
−14	34.211	19.515	1.1633	88.378	6.1537	76.175	.014149	.17126
−12	35.709	21.013	1.1169	88.139	6.6445	76.402	.015240	.17106
−10	37.256	22.560	1.0727	87.899	7.1377	76.627	.016332	.17087
− 8	38.854	24.158	1.0307	87.658	7.6333	76.852	.017425	.17067
− 6	40.504	25.808	.99066	87.416	8.1313	77.075	.018517	.17049
− 4	42.207	27.511	.95248	87.172	8.6318	77.298	.019610	.17030
− 2	43.964	29.268	.91608	86.928	9.1346	77.520	.020703	.17012
0	45.776	31.080	.88135	86.682	9.6397	77.741	.021796	.16995
2	47.644	32.948	.84821	86.435	10.147	77.960	.022889	.16978
4	49.569	34.873	.81657	86.187	10.657	78.179	.023982	.16961
6	51.552	36.856	.78636	85.937	11.169	78.397	.025075	.16944
8	53.594	38.898	.75749	85.686	11.684	78.613	.026168	.16928

* in in. Hg. vacuum
§ in cu ft/lb
# in lb/cu ft
** in Btu/lb
§§ in Btu/lb × °R

continued

continued

R-502—PROPERTIES OF SATURATED LIQUID AND SATURATED VAPOR

Temp. °F	Pressure		Volume§	Density#	Enthalpy**		Entropy§§	
	psia	psig	Vapor	Liquid	Liquid	Vapor	Liquid	Vapor
10	55.697	41.001	.72989	85.434	12.200	78.828	.027261	.16912
15	61.226	46.530	.66605	84.798	13.502	79.362	.029992	.16874
20	67.155	52.459	.60884	84.153	14.818	79.887	.032722	.16838
25	73.503	58.807	.55746	83.498	16.147	80.405	.035450	.16803
30	80.287	65.591	.51121	82.833	17.490	80.913	.038176	.16770
35	87.523	72.827	.46948	82.157	18.846	81.413	.040897	.16738
40	95.229	80.533	.43175	81.470	20.216	81.903	.043617	.16707
45	103.42	88.726	.39758	80.771	21.597	82.383	.046331	.16678
50	112.12	97.425	.36656	80.059	22.991	82.852	.049040	.16649
55	121.34	106.65	.33834	79.333	24.397	83.310	.051743	.16621
60	131.10	116.41	.31264	78.592	25.814	83.755	.054440	.16594
65	141.42	126.73	.28916	77.835	27.244	84.187	.057131	.16566
70	152.32	137.63	.26769	77.062	28.685	84.606	.059816	.16539
75	163.81	149.12	.24802	76.269	30.138	85.009	.062494	.16512
80	175.92	161.23	.22995	75.457	31.602	85.397	.065165	.16484
85	188.66	173.97	.21333	74.623	33.078	85.767	.067829	.16456
90	202.06	187.36	.19802	73.765	34.566	86.118	.070487	.16427
95	216.13	201.43	.18388	72.881	36.066	86.449	.073139	.16397
100	230.89	216.20	.17079	71.968	37.578	86.758	.075786	.16366
105	246.38	231.68	.15866	71.023	39.104	87.402	.078429	.16332
110	262.61	247.92	.14740	70.042	40.644	87.298	.081070	.16297
115	279.61	264.92	.13691	69.022	42.201	87.524	.083711	.16258
120	297.41	282.72	.12711	67.956	43.774	87.716	.086354	.16216
125	316.05	301.35	.11795	66.839	45.369	87.869	.089005	.16170
130	335.54	320.85	.10935	65.661	46.987	87.977	.091667	.16118
135	355.94	341.25	.10125	64.414	48.634	88.032	.094350	.16060
140	377.30	362.60	.093586	63.083	50.316	88.024	.097063	.15994
145	399.65	384.95	.086303	61.650	52.045	87.941	.099822	.15919
150	423.06	408.37	.079335	60.092	53.834	87.763	.10265	.15830
155	447.61	432.92	.072610	58.371	55.705	87.463	.10558	.15724
160	473.39	458.69	.066037	56.429	57.698	86.997	.10867	.15595
165	500.50	485.81	.059485	54.169	59.878	86.293	.11202	.15430
170	529.11	514.41	.052715	51.391	62.386	85.198	.11584	.15207
175	559.42	544.72	.045119	47.551	65.610	83.303	.12073	.14861
##179.9	591.0	576.3	.02857	35.00	74.81	74.81	.1346	.1346

* in in. Hg. vacuum
§ in cu ft/lb
# in lb/cu ft

** in Btu/lb
§§ in Btu/lb × °R
Critical point

ASHRAE Handbook — Fundamentals

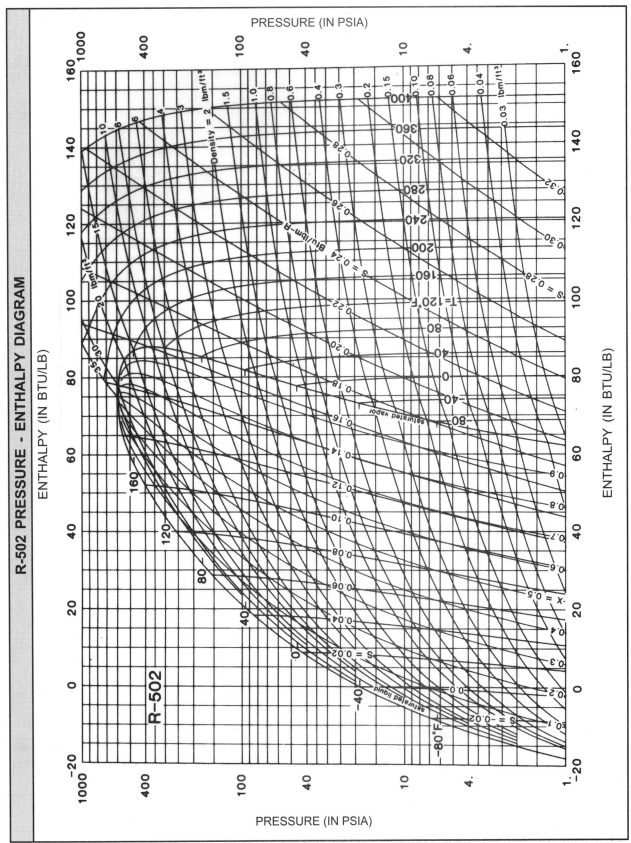

R-502 PRESSURE - ENTHALPY DIAGRAM

R-502

PRESSURE (IN PSIA)

ENTHALPY (IN BTU/LB)

R-134a PROPERTIES OF SATURATED LIQUID AND SATURATED VAPOR

Temp. °F	Vapor Pressure psia	Volume§ Vapor	Density# Liquid	Enthalpy** Liquid	Vapor	Entropy§§ Liquid	Vapor
−150	0.073	454.5455	98.86	−31.2	80.9	−0.0859	0.2761
−140	0.130	256.4103	97.91	−28.4	82.3	−0.0771	0.2693
−130	0.222	156.2500	96.96	−25.6	83.8	−0.0686	0.2633
−120	0.366	97.0874	96.01	−22.9	85.2	−0.0604	0.2579
−110	0.584	62.5000	95.06	−20.1	86.7	−0.0523	0.2531
−100	0.903	41.6667	94.11	−17.3	88.2	−0.0444	0.2488
−90	1.358	28.4091	93.15	−14.5	89.6	−0.0367	0.2450
−85	1.649	23.6407	92.68	−13.1	90.4	−0.0329	0.2433
−80	1.991	19.8413	92.20	−11.6	91.1	−0.0291	0.2416
−75	2.389	16.7224	91.72	−10.2	91.9	−0.0254	0.2401
−70	2.850	14.1844	91.24	−8.8	92.7	−0.0217	0.2386
−65	3.384	12.0773	90.76	−7.3	93.4	−0.0180	0.2373
−60	3.996	10.3306	90.27	−5.9	94.2	−0.0143	0.2360
−55	4.696	8.8810	89.78	−4.4	94.9	−0.0107	0.2348
−50	5.439	7.6687	89.30	−3.0	95.7	−0.0071	0.2336
−45	6.397	6.6489	88.81	−1.5	96.4	−0.0035	0.2326
−40	7.417	5.7904	88.31	0.0	97.2	0.0	0.2316
−38	7.860	5.4825	88.11	0.6	97.5	0.0014	0.2312
−36	8.325	5.1948	87.91	1.2	97.8	0.0028	0.2308
−34	8.811	4.9261	87.71	1.8	98.1	0.0042	0.2304
−32	9.319	4.6729	87.51	2.4	98.4	0.0056	0.2301
−30	9.851	4.4366	87.31	3.0	98.7	0.0070	0.2297
−28	10.407	4.2141	87.11	3.6	99.0	0.0084	0.2294
−26	10.987	4.0032	86.91	4.2	99.3	0.0098	0.2291
−24	11.594	3.8066	86.71	4.8	99.6	0.0112	0.2288
−22	12.226	3.6206	86.51	5.4	99.9	0.0126	0.2284
−20	12.885	3.4471	86.30	6.0	100.2	0.0139	0.2281
−18	13.572	3.2819	86.10	6.6	100.5	0.0153	0.2278
−16	14.289	3.1270	85.89	7.2	100.8	0.0167	0.2276
−14	15.035	2.9806	85.69	7.8	101.1	0.0180	0.2273
−12	15.812	2.8417	85.48	8.4	101.4	0.0194	0.2270
−10	16.620	2.7115	85.28	9.0	101.7	0.0208	0.2267
−8	17.461	2.5880	85.07	9.7	102.0	0.0221	0.2265
−6	18.334	2.4716	84.86	10.3	102.3	0.0235	0.2262
−4	19.242	2.3613	84.65	10.9	102.5	0.0248	0.2260

* in vacuum
§ in cu ft/lb
# in lb/cu ft
** in Btu/lb
§§ in Btu/lb × °R

continued

continued

	R-134a PROPERTIES OF SATURATED LIQUID AND SATURATED VAPOR						
Temp. °F	Vapor Pressure psia	Volume§ Vapor	Density# Liquid	Enthalpy** Liquid	Enthalpy** Vapor	Entropy§§ Liquid	Entropy§§ Vapor
−2	20.184	2.2568	84.44	11.5	102.8	0.0262	0.2258
0	21.163	2.1580	84.23	12.1	103.1	0.0275	0.2255
2	22.178	2.0644	84.02	12.7	103.4	0.0288	0.2253
4	23.231	1.9755	83.81	13.4	103.7	0.0302	0.2251
6	24.322	1.8911	83.60	14.0	104.0	0.0315	0.2249
8	25.454	1.8113	83.38	14.6	104.3	0.0328	0.2247
10	26.625	1.7355	83.17	15.2	104.6	0.0342	0.2244
12	27.839	1.6633	82.95	15.9	104.9	0.0355	0.2243
14	29.095	1.5949	82.73	16.5	105.2	0.0368	0.2241
16	30.395	1.5298	82.52	17.1	105.5	0.0381	0.2239
18	31.739	1.4678	82.30	17.7	105.7	0.0395	0.2237
20	33.129	1.4090	82.08	18.4	106.0	0.0408	0.2235
22	34.566	1.3528	81.86	19.0	106.3	0.0421	0.2233
24	36.050	1.2995	81.64	19.6	106.6	0.0434	0.2232
26	37.583	1.2486	81.41	20.3	106.9	0.0447	0.2230
28	39.166	1.2000	81.19	20.9	107.2	0.0460	0.2229
30	40.800	1.1538	80.96	21.6	107.4	0.0473	0.2227
32	42.486	1.1098	80.74	22.2	107.7	0.0486	0.2226
34	44.225	1.0676	80.51	22.8	108.0	0.0499	0.2224
36	46.018	1.0274	80.28	23.5	108.3	0.0512	0.2223
38	47.866	0.9890	80.05	24.1	108.6	0.0525	0.2221
40	49.771	0.9523	79.82	24.8	108.8	0.0538	0.2220
45	54.787	0.8675	79.24	26.4	109.5	0.0570	0.2217
50	60.180	0.7916	78.64	28.0	110.2	0.0602	0.2214
55	65.963	0.7234	78.04	29.7	110.9	0.0634	0.2211
60	72.167	0.6622	77.43	31.4	111.5	0.0666	0.2208
65	78.803	0.6069	76.81	33.0	112.2	0.0698	0.2206
70	85.890	0.5570	76.18	34.7	112.8	0.0729	0.2203
75	93.447	0.5119	75.54	36.4	113.4	0.0761	0.2201
80	101.494	0.4709	74.89	38.1	114.0	0.0792	0.2199
85	110.050	0.4337	74.22	39.9	114.6	0.0824	0.2196
90	119.380	0.3999	73.54	41.6	115.2	0.0855	0.2194
100	138.996	0.3408	72.13	45.1	116.3	0.0918	0.2190
110	161.227	0.2912	70.66	48.7	117.4	0.0981	0.2185
120	186.023	0.2494	69.10	52.4	118.3	0.1043	0.2181

* in vacuum
§ in cu ft/lb
# in lb/cu ft
** in Btu/lb
§§ in Btu/lb × °R

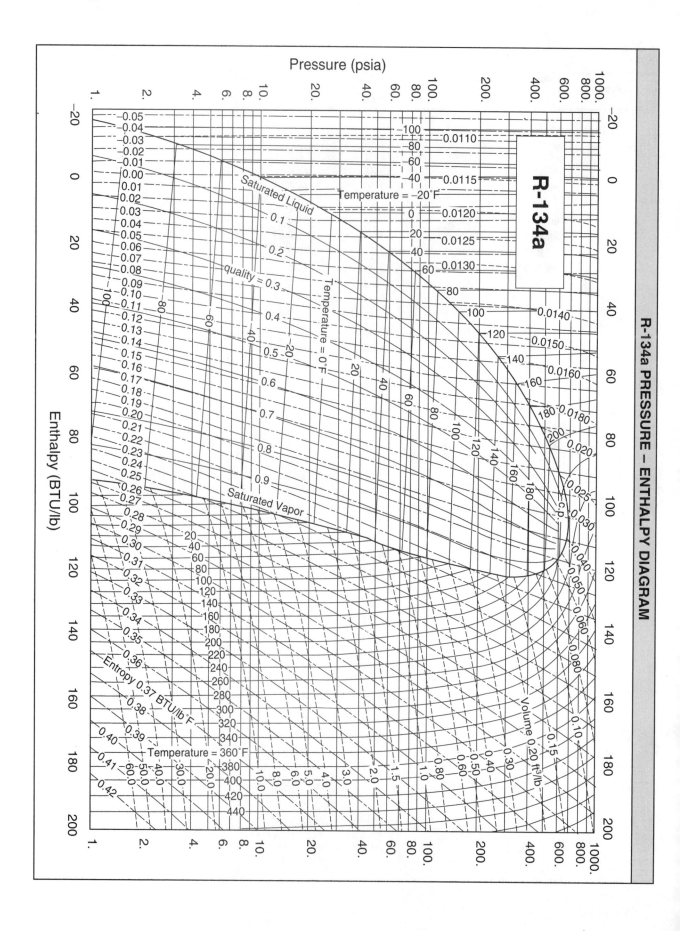

R-134a PRESSURE – ENTHALPY DIAGRAM

R-407c PROPERTIES OF SATURATED LIQUID AND SATURATED VAPOR

Temp. °F	Vapor Pressure psia	Volume§ Vapor	Density# Liquid	Enthalpy** Liquid	Enthalpy** Vapor	Entropy§§ Liquid	Entropy§§ Vapor
−150	0.15*	250.0000	99.10	−31.0	87.8	−0.0852	0.3073
−140	0.26*	151.5152	97.94	−28.3	89.2	−0.0767	0.2993
−130	0.44*	93.4579	96.78	−25.6	90.7	−0.0684	0.2921
−120	0.71*	59.5238	95.62	−22.9	92.2	−0.0603	0.2957
−110	1.10*	39.3701	94.45	−20.1	93.7	−0.0524	0.2799
−100	1.66*	26.7380	93.27	−17.4	95.2	−0.0445	0.2747
−90	2.45*	18.6220	92.09	−14.6	96.8	−0.0368	0.2701
−85	2.95*	15.6740	91.50	−13.1	97.5	−0.0331	0.2679
−80	3.53*	13.2802	90.90	−11.7	98.3	−0.0293	0.2659
−75	4.19*	11.2994	90.31	−10.3	99.1	−0.0255	0.2639
−70	4.96*	9.6525	89.71	−8.9	99.9	−0.0218	0.2621
−65	5.84*	8.2919	89.11	−7.4	100.6	−0.0181	0.2604
−60	6.84*	7.1531	88.51	−5.9	101.4	−0.0145	0.2587
−55	7.98*	6.1996	87.90	−4.5	102.2	−0.0108	0.2572
−50	9.26*	5.3937	87.30	−3.0	103.0	−0.0072	0.2557
−45	10.71*	4.7103	86.69	−1.5	103.7	−0.0036	0.2543
−40	12.33*	4.1288	86.08	0.0	104.5	0.0	0.2530
−38	13.03*	3.9200	85.84	0.6	104.8	0.0014	0.2525
−36	13.76*	3.7244	85.59	1.2	105.0	0.0029	0.2520
−34	14.53	3.5411	85.35	1.8	105.4	0.0043	0.2515
−32	15.32	3.3681	85.10	2.6	105.7	0.0060	0.2510
−30	16.16	3.2051	84.86	3.2	106.0	0.0074	0.2505
−28	17.02	3.0516	84.61	3.8	106.3	0.0088	0.2501
−26	17.93	2.9070	84.37	4.4	106.6	0.0102	0.2496
−24	18.87	2.7701	84.12	5.0	106.9	0.0116	0.2492
−22	19.85	2.6413	83.87	5.6	107.2	0.0130	0.2487
−20	20.87	2.5195	83.62	6.2	107.6	0.0144	0.2483
−18	21.94	2.4050	83.37	6.8	107.9	0.0158	0.2479
−16	23.04	2.2962	83.13	7.3	108.2	0.0168	0.2475
−14	24.19	2.1935	82.88	7.9	108.4	0.0183	0.2471
−12	25.38	2.0960	82.63	8.6	108.7	0.0197	0.2467
−10	26.62	2.0036	82.38	9.2	109.0	0.0211	0.2463
−8	27.91	1.9164	82.13	9.8	109.3	0.0225	0.2459
−6	29.24	1.8335	81.87	10.5	109.6	0.0239	0.2456
−4	30.63	1.7550	81.62	11.1	109.9	0.0253	0.2452

* in vacuum
§ in cu ft/lb
# in lb/cu ft
** in Btu/lb
§§ in Btu/lb × °R

continued

continued

R-407c PROPERTIES OF SATURATED LIQUID AND SATURATED VAPOR

Temp. °F	Vapor Pressure psia	Volume§ Vapor	Density# Liquid	Enthalpy** Liquid	Enthalpy** Vapor	Entropy§§ Liquid	Entropy§§ Vapor
–2	32.06	1.6804	81.37	11.7	110.2	0.0266	0.2448
0	33.55	1.6098	81.12	12.4	110.5	0.0280	0.2445
2	35.09	1.5425	80.86	13.0	110.8	0.0294	0.2441
4	36.69	1.4786	80.61	13.7	111.1	0.0308	0.2438
6	38.34	1.4180	80.36	14.3	111.4	0.0322	0.2435
8	40.05	1.3604	80.10	15.0	111.7	0.0336	0.2432
10	41.82	1.3053	79.85	15.6	112.0	0.0350	0.2428
12	43.64	1.2531	79.59	16.2	112.2	0.0363	0.2425
14	45.53	1.2032	79.33	16.9	112.5	0.0376	0.2422
16	47.49	1.1558	79.07	17.6	112.8	0.0390	0.2419
18	49.51	1.1105	78.82	18.2	113.1	0.0404	0.2416
20	51.59	1.0674	78.56	18.9	113.4	0.0418	0.2413
22	53.74	1.0261	78.30	19.6	113.6	0.0432	0.2410
24	55.96	0.9868	78.04	20.2	113.9	0.0446	0.2407
26	58.25	0.9493	77.78	20.9	114.2	0.0460	0.2404
28	60.61	0.9134	77.51	21.6	114.5	0.0474	0.2400
30	63.05	0.8792	77.25	22.3	114.7	0.0487	0.2399
32	65.56	0.8465	76.99	23.0	115.0	0.0501	0.2396
34	68.15	0.8151	76.72	23.7	115.3	0.0515	0.2393
36	70.81	0.7852	76.46	24.4	115.5	0.0529	0.2391
38	73.55	0.7565	76.19	25.1	115.8	0.0543	0.2388
40	76.38	0.7291	75.93	25.8	116.1	0.0557	0.2386
45	83.81	0.6656	75.26	27.5	116.7	0.0591	0.2379
50	91.78	0.6082	74.58	29.3	117.3	0.0626	0.2373
55	100.31	0.5566	73.89	31.1	117.9	0.0661	0.2367
60	109.45	0.5099	73.20	32.9	118.5	0.0696	0.2361
65	119.21	0.4677	72.51	34.8	119.1	0.0731	0.2355
70	129.61	0.4294	71.80	36.7	119.6	0.0766	0.2349
75	140.70	0.3946	71.08	38.6	120.1	0.0801	0.2343
80	152.50	0.3629	70.36	40.5	120.7	0.0836	0.2337
85	165.04	0.3340	69.62	42.5	121.1	0.0872	0.2331
90	178.36	0.3076	68.88	44.5	121.6	0.0907	0.2324
100	207.44	0.2612	67.34	48.6	122.4	0.0980	0.2312
110	240.01	0.2220	65.74	52.8	123.0	0.1053	0.2298
120	276.37	0.1887	64.06	57.3	123.5	0.1129	0.2282

* in vacuum
§ in cu ft/lb
# in lb/cu ft
** in Btu/lb
§§ in Btu/lb × °R

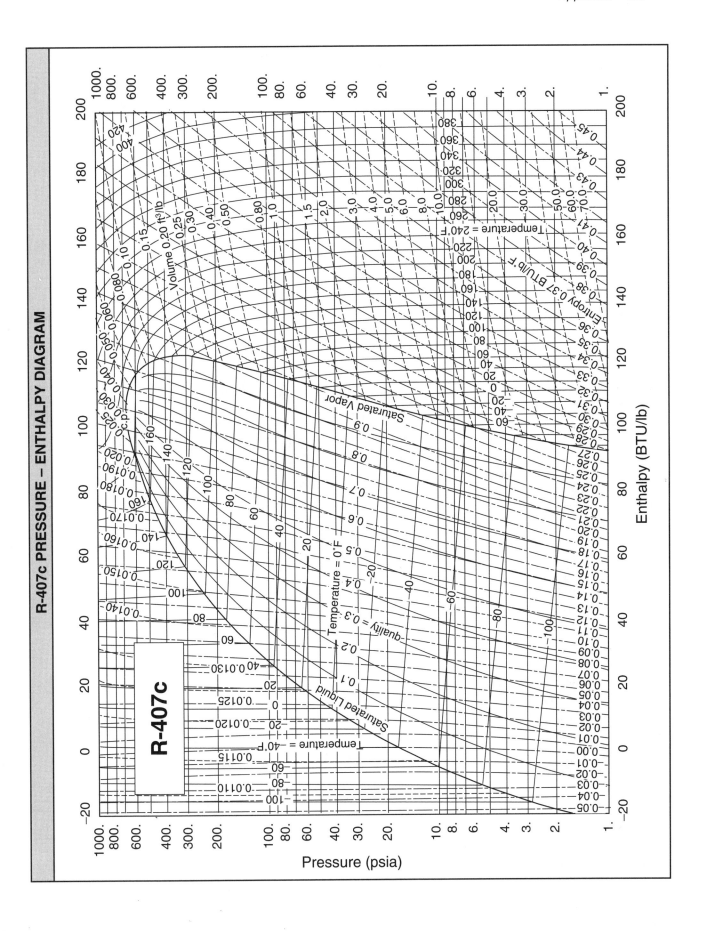

R-407c PRESSURE – ENTHALPY DIAGRAM

R-407c

Enthalpy (BTU/lb)

Pressure (psia)

HEAT TRANSFER FACTORS

Item	Design Temperature Difference*														
	30	35	40	45	50	55	60	65	70	75	80	85	90	95	100
Windows§ — Wood or Metal Frame															
Double-hung, Horizontal-slide, Casement, or Awning															
Single glass	45	50	60	65	75	80	90	95	105	110	120	125	135	140	150
With double glass or insulating glass	30	35	40	45	50	55	60	65	70	75	80	80	85	90	95
With storm sash	25	30	35	40	45	50	55	60	60	65	70	75	80	85	90
Fixed or Picture															
Single glass	40	50	55	60	70	75	85	90	95	105	110	115	125	130	140
Double glass or with storm sash	25	30	35	40	45	45	50	55	60	65	70	75	75	80	85
Jalousie															
Single glass	225	265	300	340	375	415	450	490	525	565	600	640	675	715	750
With storm sash	65	75	90	100	110	120	135	145	155	165	175	190	200	210	220
Doors§															
Sliding Glass Doors															
Single glass	75	85	100	115	125	140	150	165	175	190	200	210	225	240	250
Double glass	60	70	80	90	100	110	120	130	140	150	160	170	180	190	200
Other Doors															
Weatherstripped and with storm door	40	45	55	60	65	75	80	85	90	100	105	110	120	125	130
Weatherstripped or with storm door	70	85	95	110	120	135	145	155	170	180	195	205	215	230	240
No weatherstripping or storm door	135	160	180	200	225	250	270	290	315	340	360	380	405	430	450
Walls and Partitions§ — Wood Frame with Sheathing and Siding															
No insulation	8	9	10	11	13	14	15	16	18	19	20	21	23	24	25
R-5 polystyrene sheathing	3	4	4	5	6	6	7	7	8	8	9	9	10	10	11
R-7 batt insulation (2″ – 2¾″)	3	4	4	5	5	6	6	7	7	8	8	9	9	10	10
R-11 batt insulation (3″ – 3½″)	2	2	3	3	4	4	4	5	5	5	6	6	6	7	7
Partition Between Conditioned and Unconditioned Spaces															
Finished one side only, no insulation	17	19	22	25	28	30	33	36	39	41	44	47	49	52	55
Finished both sides, no insulation	9	11	12	14	16	17	19	20	22	23	25	26	28	29	31
Partition with 1″ polystyrene board R-5	4	4	5	5	6	7	7	8	8	9	10	10	11	11	12
R-7 insulation finished both sides	3	4	4	5	5	6	6	7	7	8	8	9	9	10	10
R-11 insulation finished both sides	2	3	3	4	4	4	5	5	6	6	6	7	7	8	8
Solid Masonry, Block, or Brick															
Plastered or plain	14	16	18	20	22	25	27	29	32	34	36	38	40	43	45
Furred, no insulation	9	10	12	13	14	16	17	19	20	22	23	25	26	28	29
Furred, with R-5 insulation	4	5	5	6	6	7	8	8	9	10	10	11	12	12	13
Basement or Crawl Space															
Above grade, no insulation	15	18	20	23	26	28	31	33	36	38	41	43	46	48	51
R-3.57 insulation (molded bead bd.)	5	6	7	8	9	10	11	12	13	14	14	15	16	17	18
R-5 insulation (ext. polystrene bd.)	4	5	6	6	7	8	9	9	10	11	12	12	13	14	14

* in °F
§ in Btu/hr per sq ft (Factors include heat loss for transmission and infiltration.)
Note: R values on this chart refer to thermal resistance value.

continued

continued

HEAT TRANSFER FACTORS															
Item	Design Temperature Difference*														
	30	35	40	45	50	55	60	65	70	75	80	85	90	95	100
Basement or Crawl Space															
Wall of crawl space used as supply plenum, R-3.57 insulation	11	12	13	14	15	16	16	17	18	19	20	21	22	23	24
Wall of crawl space used as supply plenum, R-5 insulation	9	10	10	11	12	12	13	14	15	15	16	17	17	18	19
Below grade wall	2	2	2	3	3	3	4	4	4	5	5	5	5	6	6
Ceilings and Roofs§ — Ceiling Under Unconditioned Space or Vented Roof															
No insulation	18	21	24	27	30	33	36	39	42	45	48	51	54	57	60
R-11 insulation (3″ – 3¼″)	2	3	3	4	4	4	5	5	6	6	6	7	7	8	8
R-19 insulation (5¼″ – 6½″)	2	2	2	2	2	3	3	3	4	4	4	4	4	5	5
R-22 insulation (6″ – 7″)	1	1	2	2	2	2	2	3	3	3	3	3	4	4	4
R-30 insulation	1	1	1	2	2	2	2	2	2	3	3	3	3	3	3
R-38 insulation	1	1	1	1	1	1	2	2	2	2	2	2	2	3	3
R-44 insulation	1	1	1	1	1	1	1	2	2	2	2	2	2	2	2
Roof on Exposed Beams or Rafters															
Roofing on 1½″ wood decking no ins.	10	12	14	15	17	19	20	22	24	26	27	29	31	32	34
Roofing on 1½″ wood decking 1″ insulation between roofing and decking	5	6	7	8	8	9	10	11	12	13	13	14	15	16	17
Roofing on 1½″ wood decking 1½″ insulation between roofing and decking	4	5	5	6	7	7	8	9	10	10	11	12	12	13	14
Roofing on 2″ wood plank	6	7	8	9	10	11	12	14	15	16	17	18	19	20	21
Roofing on 3″ wood plank	5	5	6	7	8	8	9	10	11	11	12	13	14	14	15
Roofing on 1½″ fiberboard decking	6	7	8	9	10	10	11	12	13	14	15	16	17	18	19
Roofing on 2″ fiberboard decking	4	5	6	7	8	8	9	10	11	11	12	13	14	14	15
Roofing on 3″ fiberboard decking	3	4	4	5	6	6	7	7	8	8	9	9	10	10	11
Roofing-Ceiling Combination															
No insulation	9	11	12	14	16	17	19	20	22	23	25	26	28	29	31
R-11 insulation	2	2	3	3	4	4	4	5	5	5	6	6	6	7	7
R-19 insulation	1	2	2	2	2	3	3	3	3	3	4	4	4	4	5
R-22 insulation (6″ – 7″)	1	1	2	2	2	2	2	3	3	3	3	3	4	4	4
Floors§ — Floors Over Unconditioned Space															
Over unconditioned room	4	5	6	6	7	8	8	9	10	11	11	12	13	13	14
No insulation	8	10	11	13	14	15	17	18	20	21	22	24	25	27	28
R-7 insulation (2″ – 2¾″)	3	3	4	4	5	5	6	6	7	7	8	8	9	9	9
R-11 insulation (3″ – 3½″)	2	2	3	3	4	4	4	5	5	5	6	6	6	7	7
R-19 insulation (5¼″ – 6½″)	1	2	2	2	2	2	3	3	3	3	4	4	4	4	4
Floor of Room Over Heated Crawl Space															
Less than 18″ below grade	35	40	40	45	45	50	50	55	55	60	60	65	65	70	75
18″ or more below grade	15	20	20	25	25	30	30	35	35	40	40	45	45	50	50

* in °F

§ in Btu/hr per sq ft (Factors include heat loss for transmission and infiltration.)

Note: R values on this chart refer to thermal resistance value.

Air Conditioning Contractors of America

COOLING HEAT TRANSFER FACTORS			
Type of Construction	**Cooling Factor***		
	15°	**20°**	**25°**
Walls			
Wood Frame with Sheeting, Siding, and Veneer or Other Finish			
No insulation, ½" gypsum board	5.0	6.4	7.8
R-11 cavity insulation + ½" gypsum board	1.7	2.1	2.6
R-13 cavity insulation + ½" gypsum board	1.5	1.9	2.3
R-13 cavity insulation + ¾" bead board (R-2.7)	1.3	1.7	2.0
R-19 cavity insulation + ½" gypsum board	1.1	1.4	1.7
R-19 cavity insulation + ¾" extruded poly	0.9	1.2	1.4
Masonry			
Above grade – No insulation	5.8	8.3	10.9
Above grade + R-5	1.6	2.3	3.1
Above grade + R-11	0.9	1.3	1.6
Below grade – No insulation	0.0	0.0	0.0
Below grade + R-5	0.0	0.0	0.0
Below grade + R-11	0.0	0.0	0.0
Ceilings			
No insulation	17.0	19.2	21.4
2" – 2½" insulation R-7	4.4	4.9	5.5
3" – 3½" insulation R-11	3.2	3.7	4.1
5¼" – 6½" insulation R-19	2.1	2.3	2.6
6" – 7" insulation R-22	1.9	2.1	2.4
10" – 12" insulation R-38	1.0	1.1	1.3
12" – 13" insulation R-44	0.9	1.0	1.1
Cathedral type (roof/ceiling combination)			
No insulation	11.2	12.6	14.1
R-11	2.8	3.2	3.5
R-19	1.9	2.2	2.4
R-22	1.8	2.0	2.2
Floors			
Over Unconditioned Space			
Over basement or enclosed crawl space (not vented)	0.0	0.0	0.0
Over vented space or garage	3.9	5.8	7.7
Over vented space or garage + R-11 insulation	0.8	1.3	1.7
Over vented space or garage + R-19 insulation	0.5	0.8	1.1
Basement Concrete Slab Floor Unheated			
No edge insulation	0.0	0.0	0.0
1" edge insulation R-5	0.0	0.0	0.0
2" edge insulation R-9	0.0	0.0	0.0
Basement Concrete Slab Floor Duct in Slab			
No edge insulation	0.0	0.0	0.0
1" edge insulation R-5	0.0	0.0	0.0
2" edge insulation R-9	0.0	0.0	0.0

* in °F
Note: R values on this chart refer to thermal resistance value.

Air Conditioning Contractors of America

DUCT HEAT LOSS MULTIPLIERS

Duct Location and Insulation Value Exposed to Outdoor Ambient Air — Attic, Garage, Exterior Wall, Open Crawl Space	Duct Loss Multipliers	
	Winter Design Below 15° F	Winter Design Above 15° F
None	1.30	1.25
R-2	1.20	1.15
R-4	1.15	1.10
R-6	1.10	1.05
Enclosed in Unheated Space — Vented or Unvented Crawl Space or Basement		
None	1.20	1.15
R-2	1.15	1.10
R-4	1.10	1.05
R-6	1.05	1.00
Duct Buried in or Under Concrete Slab — Edge Insulation		
None	1.25	1.20
R value = 3 to 4	1.15	1.10
R value = 5 to 7	1.10	1.05
R value = 7 to 9	1.05	1.00

DUCT HEAT GAIN MULTIPLIERS

Duct Location and Insulation Value Exposed to Outdoor Ambient Air — Attic, Garage, Exterior Wall, Open Crawl Space	Duct Gain Multiplier
None	1.30
R-2	1.20
R-4	1.15
R-6	1.10
Enclosed in Unconditioned Space — Vented or Unvented Crawl Space or Basement	
None	1.15
R-2	1.10
R-4	1.05
R-6	1.00
Duct Buried in or Under Concrete Slab — Edge Insulation	
None	1.10
R value = 3 to 4	1.05
R value = 5 to 7	1.00
R value = 7 to 9	1.00

COOLING HEAT TRANSFER FACTORS — WINDOWS AND DOORS*

	Single Glass			Double Glass			Triple Glass		
	Temperature Difference			Temperature Difference			Temperature Difference		
Exposure	15°	20°	25°	15°	20°	25°	15°	20°	25°
N	18	22	26	14	16	18	11	12	13
NE & NW	37	41	46	31	33	35	26	27	28
E & W	52	56	60	44	46	48	38	39	40
SE & SW	45	49	53	39	41	43	33	34	35
S	28	32	36	23	25	27	19	20	21
Wood	8.6	10.9	13.2	8.6	10.9	13.2	8.6	10.9	13.2
Metal	3.5	4.5	5.4	3.5	4.5	5.4	3.5	4.5	5.4

* Inside shading by venetian blinds or draperies.   *Air Conditioning Contractors of America*

OUTDOOR DESIGN TEMPERATURE

State and City	Lat.§	Winter DB*	Summer DB*	Daily Range*	WB*
ALABAMA					
Alexander City	32	18	96	21	79
Auburn	32	18	96	21	79
Birmingham	33	17	96	21	78
Huntsville	34	11	95	23	78
Mobile	30	25	95	18	80
Montgomery	32	22	96	21	79
Talladega	33	18	97	21	79
Tuscaloosa	33	20	98	22	79
ALASKA					
Anchorage	61	−23	71	15	60
Barrow	71	−45	57	12	54
Fairbanks	64	−51	82	24	64
Juneau	58	− 4	74	15	61
Kodiak	57	10	69	10	60
Nome	64	−31	66	10	58
ARIZONA					
Flagstaff	35	− 2	84	31	61
Phoenix	33	31	109	27	76
Tucson	32	28	104	26	72
Yuma	32	36	111	27	79
ARKANSAS					
Fort Smith	35	12	101	24	80
Hot Springs	34	17	101	22	80
Little Rock	34	15	99	22	80
Pine Bluff	34	16	100	22	81
CALIFORNIA					
Bakersfield	35	30	104	32	73
Burbank	34	37	95	25	71
Fresno	36	28	102	34	72
Laguna Beach	33	41	83	18	70
Long Beach	33	41	83	22	70
Los Angeles	33	41	83	15	70
Monterey	36	35	75	20	64
Napa	38	30	100	30	71
Oakland	37	34	85	19	66
Oceanside	33	41	83	13	70
Palm Springs	33	33	112	35	76
Pasadena	34	32	98	29	73
Sacramento	38	30	101	36	72
San Diego	32	42	83	12	71
San Fernando	34	37	95	38	71
San Francisco	37	38	74	14	64
San Jose	37	34	85	26	68
Santa Barbara	34	34	81	24	68
Santa Cruz	36	35	75	28	64
Santa Monica	34	41	83	16	70
Stockton	37	28	100	37	71
COLORADO					
Boulder	40	2	93	27	64

OUTDOOR DESIGN TEMPERATURE

State and City	Lat.§	Winter DB*	Summer DB*	Daily Range*	WB*
Colorado Spgs.	38	− 3	91	30	63
Denver	39	− 5	93	28	64
Pueblo	38	− 7	97	31	67
CONNECTICUT					
Bridgeport	41	6	86	18	75
Hartford	41	3	91	22	77
New Haven	41	3	88	17	76
Waterbury	41	− 4	88	21	75
DELAWARE					
Dover	39	11	92	18	79
Wilmington	39	10	92	20	77
DISTRICT OF COLUMBIA					
Washington	38	14	93	18	78
FLORIDA					
Cape Kennedy	28	35	90	15	80
Daytona Beach	29	32	92	15	80
Fort Lauderdale	26	42	92	15	80
Key West	24	55	90	9	80
Miami	25	44	91	15	79
Orlando	28	35	94	17	79
Pensacola	30	25	94	14	80
St. Petersburg	27	36	92	16	79
Sarasota	27	39	93	17	79
Tallahassee	30	27	94	19	79
Tampa	27	36	92	17	79
GEORGIA					
Athens	33	18	94	21	78
Atlanta	33	17	94	19	77
Augusta	33	20	97	19	80
Griffin	33	18	93	21	78
Macon	32	21	96	22	79
Savannah	32	24	96	20	80
HAWAII					
Honolulu	21	62	87	12	76
Kaneohe Bay	21	65	85	12	76
IDAHO					
Boise	43	3	96	31	68
Idaho Falls	43	−11	89	38	65
Lewiston	46	− 1	96	32	67
Twin Falls	42	− 3	99	34	64
ILLINOIS					
Aurora	41	− 6	93	20	79
Bloomington	40	− 6	92	21	78
Carbondale	37	2	95	21	80
Champaign	40	− 3	95	21	78
Chicago	41	− 9	90	15	79
Galesburg	40	− 7	93	22	78
Joliet	41	− 5	93	20	78
Kankakee	41	− 4	93	21	78
Macomb	40	− 5	95	22	79

* in °F   § in degrees

continued

continued

OUTDOOR DESIGN TEMPERATURE

State and City	Lat.§	Winter DB*	Summer DB*	Daily Range*	WB*
Peoria	40	− 8	91	22	78
Rantoul	40	− 4	94	21	78
Rockford	42	− 9	91	24	77
Springfield	39	− 3	94	21	79
Waukegan	42	− 6	92	21	78
INDIANA					
Fort Wayne	41	− 4	92	24	77
Hobart	41	− 4	91	21	77
Indianapolis	39	− 2	92	22	78
Kokomo	40	− 4	91	22	77
Lafayette	40	− 3	94	22	78
Muncie	40	− 3	92	22	76
South Bend	41	− 3	91	22	77
Terre Haute	39	− 2	95	22	79
Valparaiso	41	− 3	93	22	78
IOWA					
Ames	42	−11	93	23	78
Burlington	40	− 7	94	22	78
Cedar Rapids	41	−10	91	23	78
Des Moines	41	−10	94	23	78
Dubuque	42	−12	90	22	77
Iowa City	41	−11	92	22	80
Keokuk	40	− 5	95	22	79
Sioux City	42	−11	95	24	78
KANSAS					
Garden City	37	− 1	99	28	74
Liberal	37	2	99	28	73
Russell	38	0	101	29	78
Topeka	39	0	99	24	79
Wichita	37	3	101	23	77
KENTUCKY					
Bowling Green	35	4	94	21	79
Lexington	38	3	93	22	77
Louisville	38	5	95	23	79
LOUISIANA					
Alexandria	31	23	95	20	80
Baton Rouge	30	25	95	19	80
Lafayette	30	26	95	18	81
Monroe	32	20	99	20	79
New Orleans	29	29	93	16	81
Shreveport	32	20	99	20	79
MAINE					
Augusta	44	− 7	88	22	74
Bangor	44	−11	86	22	73
Caribou	46	−18	84	21	71
Portland	43	− 6	87	22	74
MARYLAND					
Baltimore	39	10	94	21	78
Frederick	39	8	94	22	78
Salisbury	38	12	93	18	79

* in °F   § in degrees

OUTDOOR DESIGN TEMPERATURE

State and City	Lat.§	Winter DB*	Summer DB*	Daily Range*	WB*
MASSACHUSETTS					
Boston	42	6	91	16	75
Clinton	42	− 2	90	17	75
Lawrence	42	− 6	90	22	76
Lowell	42	− 4	91	21	76
New Bedford	41	5	85	19	74
Pittsfield	42	− 8	87	23	73
Worcester	42	0	87	18	73
MICHIGAN					
Battle Creek	42	1	92	23	76
Benton Harbor	42	1	91	20	75
Detroit	42	3	91	20	76
Flint	42	− 4	90	25	76
Grand Rapids	42	1	91	24	75
Holland	42	2	88	22	75
Kalamazoo	42	1	92	23	76
Lansing	42	− 3	90	24	75
Marquette	46	−12	84	18	72
Pontiac	42	0	90	21	76
Port Huron	42	0	90	21	76
Saginaw	43	0	91	23	76
Sault Ste. Marie	46	−12	84	23	72
MINNESOTA					
Alexandria	45	−22	91	24	76
Duluth	46	−21	85	22	72
International Falls	48	−29	85	26	71
Minneapolis	44	−16	92	22	77
Rochester	43	−17	90	24	77
MISSISSIPPI					
Biloxi	30	28	94	16	82
Clarksdale	34	14	96	21	80
Jackson	32	21	97	21	79
Laurel	31	24	96	21	81
Natchez	31	23	96	21	81
Vicksburg	32	22	97	21	81
MISSOURI					
Columbia	38	− 1	97	22	78
Hannibal	39	− 2	96	22	80
Jefferson City	38	2	98	23	78
Kansas City	39	2	99	20	78
Kirksville	40	− 5	96	24	78
Moberly	39	− 2	97	23	78
St. Joseph	39	− 3	96	23	81
St. Louis	38	2	97	21	78
Springfield	37	3	96	23	78
MONTANA					
Billings	45	−15	94	31	67
Butte	45	−24	86	35	60
Great Falls	47	−21	91	28	64
Lewiston	47	−22	90	30	65

continued

continued

OUTDOOR DESIGN TEMPERATURE

State and City	Lat.§	Winter DB*	Summer DB*	Daily Range*	WB*
Missoula	46	−13	92	36	65
NEBRASKA					
Columbus	41	− 6	98	25	77
Fremont	41	− 6	98	22	78
Grand Island	40	− 8	97	28	75
Lincoln	40	− 5	99	24	78
Norfolk	41	− 8	97	30	78
North Platte	41	− 8	97	28	74
Omaha	41	− 8	94	22	78
NEVADA					
Carson City	39	4	94	42	63
Las Vegas	36	25	108	30	71
Reno	39	6	96	45	64
NEW HAMPSHIRE					
Claremont	43	− 9	89	24	74
Concord	43	− 8	90	26	74
Manchester	42	− 8	91	24	75
Portsmouth	43	− 2	89	22	75
NEW JERSEY					
Atlantic City	39	10	92	18	78
Long Branch	40	10	93	18	78
Newark	40	10	94	20	77
New Brunswick	40	6	92	19	77
Trenton	40	11	91	19	78
NEW MEXICO					
Albuquerque	35	12	96	27	66
Carlsbad	32	13	103	28	72
Gallup	35	0	90	32	64
Los Alamos	35	5	89	32	62
Santa Fe	35	6	90	28	63
Silver City	32	5	95	30	66
NEW YORK					
Albany	42	− 6	91	23	75
Batavia	43	1	90	22	75
Buffalo	42	2	88	21	74
Geneva	42	− 3	90	22	75
Glens Falls	43	−11	88	23	74
Ithaca	42	− 5	88	24	74
Kingston	41	− 3	91	22	76
Lockport	43	4	89	21	76
New York City	40	11	92	17	76
Niagara Falls	43	4	89	20	76
Rochester	43	1	91	22	75
Syracuse	43	− 3	90	20	75
Utica	43	−12	88	22	75
NORTH CAROLINA					
Charlotte	35	18	95	20	77
Durham	35	16	94	20	78
Greensboro	36	14	93	21	77
Jacksonville	34	20	92	18	80

OUTDOOR DESIGN TEMPERATURE

State and City	Lat.§	Winter DB*	Summer DB*	Daily Range*	WB*
Wilmington	34	23	93	18	81
Winston-Salem	36	16	94	20	76
NORTH DAKOTA					
Bismarck	46	−23	95	27	73
Fargo	46	−22	92	25	76
Grand Forks	47	−26	91	25	74
Williston	48	−25	91	25	72
OHIO					
Akron-Canton	40	1	89	21	75
Athens	39	0	95	22	78
Bowling Green	41	− 2	92	23	76
Cambridge	40	1	93	23	78
Cincinnati	39	1	92	21	77
Cleveland	41	1	91	22	76
Columbus	40	0	92	24	77
Dayton	39	− 1	91	20	76
Fremont	41	− 3	90	24	76
Marion	40	0	93	23	77
Newark	40	− 1	94	23	77
Portsmouth	38	5	95	22	78
Toledo	41	− 3	90	25	76
Warren	41	0	89	23	74
OKLAHOMA					
Bartlesville	36	6	101	23	77
Chickasha	35	10	101	24	78
Lawton	34	12	101	24	78
McAlester	34	14	99	23	77
Norman	35	9	99	24	77
Oklahoma City	35	9	100	23	78
Seminole	35	11	99	23	77
Stillwater	36	8	100	24	77
Tulsa	36	8	101	22	79
Woodward	36	6	100	26	78
OREGON					
Albany	44	18	92	31	69
Astoria	46	25	75	16	65
Baker	44	− 1	92	30	65
Eugene	44	17	92	31	69
Grants Pass	42	20	99	33	71
Klamath Falls	42	4	90	36	63
Medford	42	19	98	35	70
Portland	45	17	89	23	69
Salem	44	18	92	31	69
PENNSYLVANIA					
Allentown	40	4	92	22	76
Altoona	40	0	90	23	74
Butler	40	1	90	22	75
Erie	42	4	88	18	75
Harrisburg	40	7	94	21	77
New Castle	41	2	91	23	75

* in °F   § in degrees

continued

continued

OUTDOOR DESIGN TEMPERATURE					
		Winter	Summer		
State and City	Lat.§	DB*	DB*	Daily Range*	WB*
Philadelphia	39	10	93	21	77
Pittsburgh	40	1	89	22	74
Reading	40	9	92	19	76
West Chester	39	9	92	20	77
Williamsport	41	2	92	23	75
RHODE ISLAND					
Newport	41	5	88	16	76
Providence	41	5	89	19	75
SOUTH CAROLINA					
Charleston	32	25	94	13	81
Columbia	33	20	97	22	79
Florence	34	22	94	21	80
Sumter	33	22	95	21	79
SOUTH DAKOTA					
Aberdeen	45	−19	94	27	77
Brookings	44	−17	95	25	77
Huron	44	−18	96	28	77
Rapid City	44	−11	95	28	71
Sioux Falls	43	−15	94	24	76
TENNESSEE					
Athens	35	13	95	22	77
Chattanooga	35	13	96	22	78
Dyersburg	36	10	96	21	81
Knoxville	35	13	94	21	77
Memphis	35	13	98	21	80
Murfreesboro	34	9	97	22	78
Nashville	36	9	97	21	78
TEXAS					
Abilene	32	15	101	22	75
Alice	27	31	100	20	82
Amarillo	35	6	98	26	71
Austin	30	24	100	22	78
Beaumont	29	27	95	19	81
Big Spring	32	16	100	26	74
Brownsville	25	35	94	18	80
Corpus Christi	27	31	95	19	80
Dallas	32	18	102	20	78
El Paso	31	20	100	27	69
Fort Worth	32	17	101	22	78
Galveston	29	31	90	10	81
Houston	29	27	96	18	80
Huntsville	30	22	100	20	78
Laredo	27	32	102	23	78
Lubbock	33	10	98	26	73
Mcallen	26	35	97	21	80
Midland	31	16	100	26	73
Pecos	31	16	100	27	73
San Antonio	29	25	99	19	77
Temple	31	22	100	22	78
Tyler	32	19	99	21	80

OUTDOOR DESIGN TEMPERATURE					
		Winter	Summer		
State and City	Lat.§	DB*	DB*	Daily Range*	WB*
Victoria	28	29	98	18	82
Waco	31	21	101	22	78
Wichita Falls	33	14	103	24	77
UTAH					
Cedar City	37	− 2	93	32	65
Logan	41	− 3	93	33	65
Provo	40	1	98	32	66
Salt Lake City	40	3	97	32	66
VERMONT					
Barre	44	−16	84	23	73
Burlington	44	−12	88	23	74
Rutland	43	−13	87	23	74
VIRGINIA					
Charlottesville	38	14	94	23	77
Fredericksburg	38	10	96	21	78
Harrisonburg	38	12	93	23	75
Lynchburg	37	12	93	21	77
Norfolk	36	20	93	18	79
Petersburg	37	14	95	20	79
Richmond	37	14	95	21	79
Roanoke	37	12	93	23	75
Winchester	39	6	93	21	77
WASHINGTON					
Aberdeen	46	25	80	16	65
Bellingham	48	10	81	19	68
Olympia	46	16	87	32	67
Seattle-Tacoma	47	17	84	22	66
Spokane	47	− 6	93	28	65
Walla Walla	46	0	97	27	69
WEST VIRGINIA					
Charleston	38	7	92	20	76
Clarksburg	39	6	92	21	76
Parkersburg	39	7	93	21	77
Wheeling	40	1	89	21	74
WISCONSIN					
Beloit	42	− 7	92	24	78
Fon Du Lac	43	−12	89	23	76
Green Bay	44	−13	88	23	76
La Crosse	43	−13	91	22	77
Madison	43	−11	91	22	77
Milwaukee	42	− 8	90	21	76
Racine	42	− 6	91	21	77
Sheboygan	43	−10	89	20	77
Wausau	44	−16	91	23	76
WYOMING					
Casper	42	−11	92	31	63
Cheyenne	41	− 9	89	30	63
Laramie	41	−14	84	28	61
Rock Springs	41	− 9	86	32	59
Sheridan	44	−14	94	32	66

* in °F   § in degrees

Air Conditioning Contractors of America

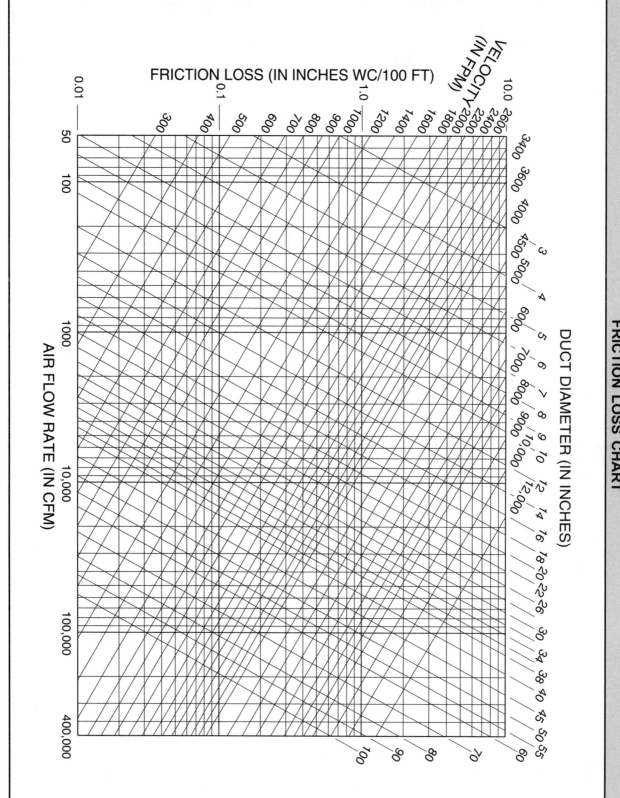

VELOCITY
(IN FPM)

FRICTION LOSS (IN INCHES WC/100 FT)

DUCT DIAMETER (IN INCHES)

AIR FLOW RATE (IN CFM)

FRICTION LOSS CHART

HEATING TROUBLESHOOTING CHART

Complaint	Symptom	Cause	Description	Section, Component, or Part
No heating	Cool air out of supply registers	Controls	Burner not operating	Transformer Thermostat Heating relay Limit control
		Fuel supply	Burner not operating	Tanks empty Valves shut off
		Combustion controls	Burner not operating	Thermocouple Pilot safety valve Fuel valve
	No air out of registers	Power controls	No electricity	Disconnect switches Fuses blown Circuit breaker open
		Circulation system	Blower not operating	Blower motor Blower relay Blower belts
Insufficient heat	Air out of supply registers too cool	Controls	Burner not operating	Control transformer Thermostat Heating relay Limit control
		Combustion controls		Thermocouple Pilot safety valve Fuel valve
			Burner operating intermittently	Thermostat Limit control
Too much heat	Building overheats	Controls	Temperature exceeds setpoint	Thermostat Gas valve Heating relay
Fluctuating heat	Temperature changes in building	Controls	Burner cycles OFF and ON	Limit control Blower relay
		Circulation system	Burner cycles OFF and ON	Blower motor Blower relay
		Ductwork	Air supply	Sizing Dampers
System noisy	Equipment noisy	Controls	Humming	Transformer Relays
		Blower	Squeaks or rattles	Blower wheel Motor
		Burner	Whistle	Air adjustment
		Distribution	Pops and rattles	Ductwork Registers
Odor	Smell of fuel around unit	Burner connections	Smell fuel	Fuel line Fuel pump Fuel valves Filter
	Smell flue gas	Unit	Smell flue gas at unit	Draft diverter Barometric damper Flue
			Smell flue gas at supply registers	Heat exchanger
Cost of operation	Unit	Distribution system	Unit runs too long	Dirty filters Ductwork undersized Registers too small

AIR CONDITIONING TROUBLESHOOTING CHART

Complaint	Symptom	Cause	Description	Section, Component, or Part
No cooling	No air out of registers	Electric power	No power	Disconnect Fuses Circuit breakers
		Controls	Control voltage	Transformer Thermostat Limit control
		Blower	Not operating	Blower wheel Motor
		Evaporator	Obstructed	Coil
		Air distribution	Dampers	Closed
	Warm air out of registers	System	Equipment	Undersized
		Air distribution	Ductwork	Undersized
			Blower	Wheel Motor
			Dampers	Closed
		Controls	Calibration Setting	Thermostat Pressure switches
		Compressor	Not operating	Motor Valves
		Condenser	Insufficient air	Blower Motor Dirty coil
		Evaporator	Insufficient air	Dirty coil Dirty filter
Not enough cooling	Air supply too warm	Compressor	Cycling	Pressure switches Overloads
		Condenser	Not enough air	Blower wheel Motor Dirty coil Noncondensable gases
		Evaporator	Not enough air	Blower wheel Motor Dirty coil
			Expansion device	Adjustment Faulty
		Air distribution	Ductwork	Undersized Not insulated
			Supply registers	Dampers closed Too small
			Return air grills	Location Size
			Filters	Dirty
		Equipment	Runs all the time	Sized too small
Too cool	Air is cold out of registers	Controls	Calibration Anticipation	Thermostat
	Cold drafts	Air distribution	Air pattern	Supply registers location System balance

continued

continued

AIR CONDITIONING TROUBLESHOOTING CHART				
Complaint	**Symptom**	**Cause**	**Description**	**Section, Component, or Part**
Noisy system	Noisy units	Compressor	Knocking sounds	Mechanical problems
			Liquid slugs	Expansion device Refrigerant charge
		Condenser	Blower	Wheel Motor
		Evaporator	Blower	Wheel Motor
		Refrigerant lines	Vibrations	Lines too rigidly attached to supports
	Noise out of registers	Ductwork	Popping and ticking	Ducts need bracing
Too expensive to operate	Unit runs constantly	Equipment	Runs constantly	Undersized
		Controls	Out of adjustment	Thermostat
			Welded contacts	Thermostat Compressor relay
		Air circulation	Motors run constantly	Shorted electrical circuits
			Filters	Dirty
		Evaporator	Too little air	Dirty coil Blower not operating
		Condenser	Too little air	Dirty coil Blower not operating
	Needs frequent service	System	Improperly designed	Undersized
			Improperly installed	Poor workmanship Inadequate maintenance

MAINTENANCE CHECKLIST

Company Name _____

Address _____

Homeowner's Name _____

Address _____

Date _____ Technician _____ Time In _____ Time Out _____

Furnace Make and Model _____

Comments by Homeowner _____

AT THERMOSTAT

— Record thermostat setpoint temperature. _____

— Check thermostat for dust or dirt and level.

— Turn thermostat to lowest cooling setting.

AT FURNACE

— Check supply voltage and record.
Time _____ Voltage _____

— Clean or change filters.

— Clean out blower wheel and blower compartment.

BELT-DRIVE BLOWERS

— Check wiring in blower compartment for loose connections or bad insulation.

— Remove blower belt. Check for wear.

— Check blower belt and motor bearings.

— Check pulley and drive alignment.

— Check pulley and drive set screws for tightness.

— Check motor bracket for tightness.

— Check blower for free operation.

— Lubricate blower and motor bearings.

— Put belt back on blower and drive pulleys and check belt tension slippage.

DIRECT-DRIVE BLOWERS

— Check wiring in blower cabinet for loose connections and bad insulation.

— Check motor bearings.

— Check for free blower operation.

— Check blower set screws for tightness.

— Lubricate motor bearings.

CONDENSING UNIT

— Check and clean condenser coil.

— Oil condenser fan motor.

— Check supply voltage and record.
Time _____ Voltage _____

— Check all wiring for loose connections.

— Check all wiring for damaged insulation.

— Gauge refrigeration system and check operating pressures.

— Check refrigerant charge.

— Check amperage draw on condenser fan motor.
Nameplate _____ Actual _____

— Check amperage draw on compressor.
Nameplate _____ Actual _____

— Visually inspect connecting tubing and coils for evidence of oil leak.

— Return thermostat to original setpoint temperature.

EVAPORATOR COIL

— Check and clean cooling coil.

— Check and clean condensate drain.

— Check static pressure.

— Check temperature difference over coil.

— Check for proper voltage at transformer and evaporator blower.

If unit is not running, refer to Service Handbook for cause of trouble.

BE SURE TO LEAVE ALL AREAS NEAT AND CLEAN!

Contractor: _____

Name of Job: _____

Address: _____

Date _____

By: _____

Winter: Indoor Design Temp. _____   Outdoor Design Temp. _____   Design Temperature Difference _____

Summer: Outdoor Design Temp. _____   Indoor Design Temp. _____   Design Temperature Difference _____

		1			2			3			4			5			6			Building Component Subtotals	
		Area	Btu/hr H	Btu/hr C	Area	Btu/hr H	Btu/hr C	Area	Btu/hr H	Btu/hr C	Area	Btu/hr H	Btu/hr C	Area	Btu/hr H	Btu/hr C	Area	Btu/hr H	Btu/hr C	Btu/hr H	Btu/hr C
1	Name of Room																				
2	Running Feet of Exposed Wall																				
3	Room Dimensions																				
4	Ceiling Height / Exposure																				
5	Gross Exposed Wall Area — N / S / E / W																				
6	Windows and Glass Doors (H)																				
7	Windows and Glass Doors (C) — N / E&W / S																				
8	Other Doors																				
9	Net Exposed Walls																				
10	Ceilings																				
11	Floors																				
12	People @ 300 Appliances @ 1200																				
13	Subtotal Btu/hr Loss																			Totals	
14	Sensible Btu/hr Gain																				
15	Subtotal Btu/hr Gain (1.3)																				

Contractor: _____

Name of Job: _____

Address: _____

Date _____

By: _____

Winter: Indoor Design Temp. _____   Outdoor Design Temp. _____   Design Temperature Difference _____

Summer: Outdoor Design Temp. _____   Indoor Design Temp. _____   Design Temperature Difference _____

			Factors		1			2			3			4			5			6			Building Component Subtotals
			H	C	Area	Btu/hr H	Btu/hr C	Area	Btu/hr H	Btu/hr C	Area	Btu/hr H	Btu/hr C	Area	Btu/hr H	Btu/hr C	Area	Btu/hr H	Btu/hr C	Area	Btu/hr H	Btu/hr C	
1	Name of Room																						
2	Running Feet of Exposed Wall																						
3	Room Dimensions																						
4	Ceiling Height Exposure																						
5	Gross Exposed Wall Area	Types Exposure																					
		N																					
		S																					
		E																					
		W																					
6	Windows and Glass Doors (H)																						
7	Windows and Glass Doors (C)	N																					
		E&W																					
		S																					
8	Other Doors																						
9	Net Exposed Walls																						
10	Ceilings																						
11	Floors																						
12	People @ 300 Appliances @ 1200																						
13	Subtotal Btu/hr Loss																						Totals
14	Sensible Btu/hr Gain																						
15	Subtotal Btu/hr Gain (1.3)																						